Hamlyn all-colour paperbacks

Catherine Jarman BSc

Evolution of Life

illustrated by Peter Thornley

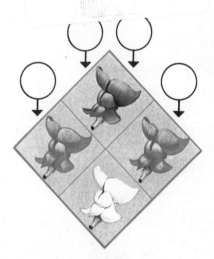

Hamlyn · London
Sun Books · Melbourne

FOREWORD

'Happy the man whose lot is to know
the secrets of the earth. He hastens not
To work his fellow's hurt by unjust deeds
But with rapt admiration contemplates
Immortal Nature's ageless harmony
And when and how her order came to be.'

Euripides

Evolution is no more than the idea that all the varied kinds of animals and plants which we know have developed from earlier types by completely natural changes during the passage of time. It is hoped that this book will make the reader consider the origin of all the plants and animals around him, and cause him to trace them back through the millions of years to their beginnings.

How evolution has occurred and is still happening is an involved concept but it is hoped that the drawings and text in this book will give the readers insight into this intriguing subject.

C.J.

Published by The Hamlyn Publishing Group Limited
London . New York . Sydney . Toronto
Hamlyn House, Feltham, Middlesex, England
In association with Sun Books Pty Melbourne

Copyright © The Hamlyn Publishing Group Limited 1970

ISBN 0 600 00093 1
Phototypeset by Filmtype Services Limited, Scarborough
Colour separations by Schwitter Limited, Zurich
Printed in England by Sir Joseph Causton & Sons Limited

CONTENTS

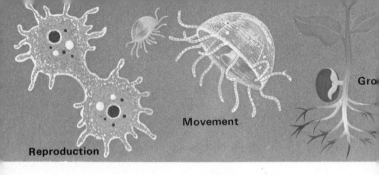

Reproduction

Movement

Gro

INTRODUCTION

The nature of life

What is life? There are many replies to this question, and man continues to search for the complete answer. However, all living things have a few basic and common characteristics.

There are processes which both plants and animals show and one can distinguish the living from the non-living after scientific investigation. The most familiar processes are growth, movement, metabolism, reproduction, irritability, feeding, respiration and excretion. To function successfully, all those complex processes must be performed by living things.

The basic 'stuff' of living organisms is *protoplasm*. There is no set composition of this and it varies between one individual and the next. It consists of a complex mixture of water, organic compounds and a number of salts. From the origin of life to the present day, all living things, no matter how diverse or unrelated, have shared these common processes that are necessary for life to continue.

Common processes of life (*above*) and diversity of forms (*below*)

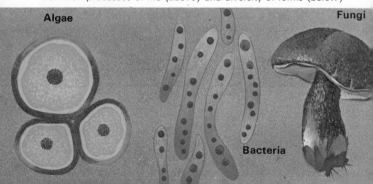

Algae

Fungi

Bacteria

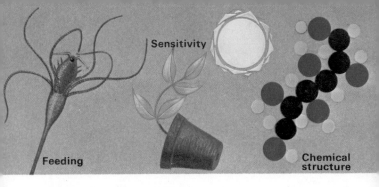

Sensitivity

Feeding

Chemical structure

The abundance and diversity of life

When the numbers of individuals of any particular type of plant or animal life are studied, the figures are quite staggering. For example, a single salmon produces as many as twenty-eight million eggs in a season. Of course, not all these eggs survive, indeed only a small fraction will have a long life. One can imagine the chaos created if all the offspring of an individual did survive.

This abundance of life is a major factor in the evolution of living things leading as Darwin showed, to natural selection. Examples of the numerous forms of 'life' show the numbers of some of the species in certain groups.

Angiosperms (flowering plants)	150,000
Thallophytes (algae, fungi, lichens)	107,000
Bryophytes (liverworts, mosses)	23,000
Pteridophytes (ferns, horsetails)	10,000
Invertebrates and Vertebrates	1,120,000

The large number of animal species is noticeable, but more than three-quarters are insects.

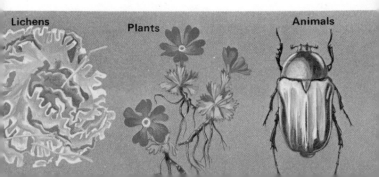

Lichens

Plants

Animals

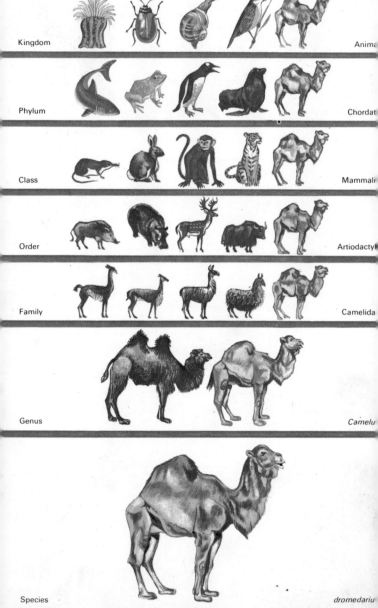

Classification of an Arabian Camel

Kingdom	Anima[l]
Phylum	Chordat[a]
Class	Mammal[ia]
Order	Artiodacty[la]
Family	Camelida[e]
Genus	*Camelu[s]*
Species	*dromedariu[s]*

Classification ✓

For centuries man has tried to arrange living organisms into some kind of logical order or grouping. This arranging in scientific groupings is called classification or taxonomy. However, it is nearly the art of the impossible, as perfect classification is nonexistent. Nevertheless, biologists strive for perfection for after all, animals and plants cannot be studied fully until they have been classified.

At first, animals were grouped according to how they were adapted for living. Thus whales were classed as fish as they lived in the sea. Similarly, bats were grouped as birds because they could fly. This classification based on how the animal was adapted, led to wrong groupings, and did not show the relationship between the various animals.

The next phase was far better, but still inaccurate by modern standards. This grouping was based on the form of the animal, known as morphology. This system was best shown by Linnaeus (1707-1778), a Swedish naturalist. It is based on the assumption that there are as many types of animals now in existence as were produced during the six days of creation. His thoughts followed the traditional assumptions of that time about the kinds of living things. Linnaeus was well aware of the anatomical resemblances between the types or species; between for example, horse and donkey, newt and lizard, ape and man. He correctly grouped lions, tigers and domestic cats into a Linnaean group, the Felidae. These animals, although they exhibited many resemblances, were merely grouped according to the overall structural plan. All species, it was assumed, remained unaltered through the course of time and a possible link between generations was not realised. This form of classification, and the assumptions which led to and from it, did not suggest evolution.

When the fact of evolution had been accepted in the second half of the 19th century, taxonomists tried to express the evolutionary relationships between animals and plants. The terminology used by Linnaeus was retained but a wider range of evidence was taken into account. Morphology was still very important, but the evidence of fossils (palaeontology), genetics, embryology, immunology and geographical distribution was also taken into consideration.

Aristotle understood the importance of change in the development of 'form'.

EVOLUTION

Some early theories

Today, evolution is accepted as an undisputed fact by the majority of people. However, it has not always been a scientific theory. After all, the notion of life evolving is not one which comes immediately and naturally to the majority of minds.

Many ideas of evolution were presented in symbolic forms in myths. Thales (640-546 BC) believed all life came from water, whereas Anaximenes of Miletus (circa 500 BC) said all things came from air. Most mythological ideas are linked with the 'elements' such as earth, fire, water and air. Some ideas were quite fantastic and comparatively recent. A. Kircher (1601-1680) thought orchids gave birth to birds and small men and B. de Maillet (1656-1738) recognized the true nature of fossils, but thought birds were derived from flying fishes, lions from sea-lions, and men from mermen.

Theological thoughts interpreted the idea of evolution as an act of God. The Hebrews accepted the religious writings of the

Bible, that living things were specially created in six days by an all-powerful personal god. Special creation was an act that was apparently done once and for all. Although the biblical account gives six 'days', this ancient term is interpreted to mean six epochs. The special creation theory is based on the analogy of the making of the universe by God, with the manufacture of articles by man. However, theological views and the evolution theory are compatible, if we consider that both have a common background in the productivity of life.

The theories of the Greek 'thinkers' give the other main thoughts on evolutionary ideas. In his biological philosophy, Aristotle (384-322 BC) showed the importance of change and the theory of 'forms' or 'essences'. He understood that the speck of matter in a hen's egg becomes, by definite stages, a chicken. Also, by numerous examples he understood complete metamorphosis where organisms such as frogs and insects, developed and matured in definite changing stages. His work shows he conceived the idea that life began as undifferentiated 'matter' with a 'potentiality' for turning into a form although

Special creation in six days is described in the Bible.

his ideas were limited to individuals. Any evolutionary thought of how one individual can somehow change into another was absent.

The act of special creation and the Aristotle philosophy of 'forms' were quite sufficient to satisfy nearly everybody during the Middle Ages and even through the Renaissance and Reformation, when one might have expected some fresh thoughts on the philosophy of biology.

Up to the 18th century the Earth was assumed to have been created in 4,004 BC, as decided by the Archbishop of Armagh (1581-1656). It was not until the science of geology rapidly advanced in the late 18th century that doubt was thrown on the 6,000-year-old Earth theory. Several doubted the few thousand years and suggested several hundred thousands more appropriate.

During the late 17th and into the 18th century a true appreciation of fossils developed. Up to this period fossils were thought to develop from moist seed-bearing vapours, blown from the seas into the crevices of the earth. Martin Lister (1638-1712) and Robert Hooke (1635-1703) suggested the possible value of distinctive fossils in relation to the strata in which they were found. Lister did not, however, think they were fossilized animal or plant remains.

No startling theories followed until 1788, when James Hutton (1726-1797), an Edinburgh doctor, published as essay

Erasmus Darwin (*left*) and Cuvier (*below*)

'A Theory of the Earth' (1795). 'Uniformitarianism' was Hutton's explanation of the past history of the Earth. By this, Hutton stated that divinely ordained floods and universal rock upheavals were unnecessary to explain the existence and nature of oceans and mountain chains. Hutton argued that the Earth's processes were gradual in their nature, rock and soil being worn away by the process of erosion. His ideas led him to state that the continents could not have been formed in a few thousand years, but millions of years. He implied that the Earth was infinitely old and had an equally vast future.

Although the Earth had altered, the changing of animals and plants was still not suggested. At the turn of the 18th century the work of William Smith (1769-1838) did a great deal in establishing the value of animal fossils in the recognition of strata in geology.

Cuvier, a zoology professor in Paris, was Smith's contemporary and had a profound knowledge of comparative anatomy which enabled him to reconstruct animal skeletons of the past from fossil bones or markings.

This great spurt in the geological sciences provided a wealth of information about animal and plant characteristics. Many species of plants and animals described by geologists were extinct, so the idea of a succession of animal types through time began to unfold. However no one crossed the threshold of an evolutionary interpretation of life. One fact that helped to prevent this was a non-evolutionary interpretation of animal life provided by Cuvier himself. This was the theory of 'catastrophism'. He concluded that each fauna had been destroyed by a gigantic catastrophe and a new fauna then migrated from distant regions. Noah's flood was taken to be the most recent catastrophe.

Erasmus Darwin (1731-1802) is most famous for being grandfather to Charles Darwin. However, his thoughts and speculations caused a sensation in his day, in England and on the continent. His notable scientific work was 'Zoonomia', an attempt to find out the organic laws of life. His natural philosophy was that spirit and matter are the foundations of nature. Life he defined as being due to a special force, irritability, and all manifestations of life are due to the contraction of fibres, induced by irritation.

Lamarck (*left*) and the classic example for his theory of evolution (*right*)

The theories of Lamarck

Lamarck (1744-1829), a friend of Cuvier, was the first person to tackle the problem of how species originated. Boldly, he proclaimed that there was no essential difference between species and varieties, that species like varieties were subject to change, and that 'transformation', not immutability, was the basis of life.

In his work 'Zoological Philosophy' (1809) Lamarck suggested that a transformation of species might have occurred by the 'inheritance of acquired characters'. He drew attention to the observation that animals do change in form during their life span. For instance, a man may build up his muscles during his life by repeated exercise. Assuming this acquired character was passed on or inherited by the offspring, then the next generation would be born with slightly stronger muscles. In his 'Zoological Philosophy' Lamarck argued that if this change was continued through several generations, it would result in a completely new species.

The giraffe is the classic example quoted. The giraffe with its long neck could have developed from an animal with quite a short neck. Straining its neck through untold generations to reach higher levels of foliage presumably resulted in an elongated neck. This was the basis of Lamarck's theory, with species gradually changing in form by the differential use of organs and faculties throughout time.

The fact that Lamarck suggested that animals exert a will to conduct their lives in a certain manner and that he cannot explain plant evolution, shows his theory had definite imperfections. He did no experiments to provide evidence to support his theory at the time, and a considerable amount of later experimenting failed to produce examples which convincingly showed any such inheritance.

Darwin – his life and theories ✓

The 'Theory of Evolution' is usually associated with Charles Darwin whose genius in the late 1850's gave the world the new idea of natural selection.

Darwin was born in Shrewsbury in 1809, the son of a wealthy doctor. At the local public school he received the usual course of Latin and Greek verses with classical geography and history. However, he became more interested in natural history and was therefore sent up to Edinburgh to study for the medical profession. The sight of two operations under the conditions of that time, before anaesthetics had been brought into use, revolted the seventeen-year-old Darwin. After two years, when it was quite apparent he was totally un-suited for medical life, he returned home to Shrewsbury. His family now decided he should study for the Church. Charles did not violently object and so 1828 found him in Cambridge. On graduating Darwin was resigned to take Church Orders when fate intervened to set an apparently unambitious man, with little distinction to his name, on the way to show the world his spark of genius and depth of reasoning. On the advice of his professor and friend, John Henslow, Darwin joined the H.M.S. Beagle to journey to South America as an unpaid naturalist.

Up to his departure from England Darwin had no cause to doubt the immutability of species. He studied Sir Charles Lyell's 'Principles of Geology' and acquired a general background of 'uniformitarianism' rather than 'catastrophism'. However, on the voyage certain observations turned Darwin against the immutability of species theory.

On his trip to the Galapagos Islands, he was amazed by the finch differences (see pages 56 to 57), and he wondered why there had been such a fantastic number of species created in the Islands. He also compared the fauna of the Galapagos with that of the Cape Verde Islands where in spite of similar physical conditions the animal species were totally different; those of the Galapagos resembling those of South America and the fauna of the latter resembles that of Africa. In his travels through South America Darwin noticed that animals of different species showed a great deal of resemblance to one another in regions where there was only slight distance between each species.

During the voyage of the survey ship 'Beagle' (1831–1836) Darwin became convinced that species evolved by natural processes.

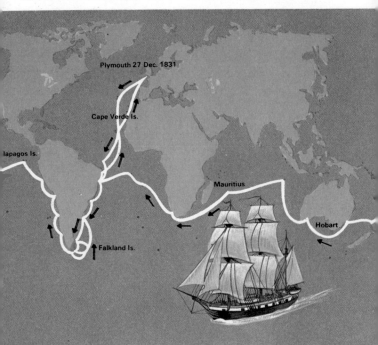

The origin of species

On his return to England in 1836, Darwin was convinced of evolution, but how these changes had taken place still had to be answered. Darwin had some ideas and wrote numerous notes but he kept them secret except to his close friends, Lyell and Hooker. The basis of his thoughts was that *species had evolved by natural processes from a few simple primordial forms or even one form.*

Darwin led a secluded life at Downe with his wife, a daughter of the Wedgwood family. He took great interest in the breeding methods of domestic animals and fancy types such as pigeons. He realized that breeders always chose certain characteristics and continued to select required variants through several generations, until the desired new breed was obtained. These bred true if mated only between themselves. This 'artificial selection' gave a first clue, but how does selection operate in the wild without any conscious being directing its course?

When Darwin read the essay 'The Principles of Population' by Thomas Robert Malthus, an English clergyman, he realised that, under intense conditions of plants and animals competing between and among themselves to live, any variations continued would be those which increased the organism's ability to leave fertile offspring. Those which decreased the animal's or plant's number would eventually be eliminated

Some of the many varieties of domestic pigeon that have evolved by artificial selection.

Rock Dove

Fantail

and thus 'natural selection' would take place. Though Darwin had the framework to his evolutionary theory in 1838 it was another twenty years before the public was to learn of his revolutionary ideas, when his hand was forced by Alfred Russel Wallace.

Wallace had read Malthus in the early 1840's but it was not until 1847 that he was convinced of evolution. In that year he joined a collecting expedition to South America and later worked in Malaya and the eastern Indian archipelago. Surrounded by evolutionary evidence Wallace wrote a short paper arguing that species must have come into existence slowly but he, like Darwin, was worried how the change took place. In 1858, while recovering from an attack of malaria, Wallace thought of Malthus and suddenly came to the same conclusion as Darwin. When recovered from his illness, Wallace quickly wrote a twenty page essay outlining the theory. He then sent this identical theory to Darwin in Kent to ask his opinion for publication.

Seeing his own thoughts written by a distant, unknown person must have shocked Darwin. He realized he had no time to lose and on the advice of his friends wrote a short paper giving his views and had the point communication read to learned naturalists at the Linnean Society in London in July 1858. Darwin followed this by publishing his book 'The Origin of Species' or 'The Preservation of the Favoured races in the Struggle for Life' in November of the following year.

Pouter

Short-faced Tumbler

17

In his book Darwin put forward four points he knew to be true and three which we now know to be true. They can be briefly summarized as follows:-

1. Plants and animals produce a far greater number of reproductive cells than ever give rise to mature individuals.
2. The number of individuals in species remain more or less constant.
3. Because of this there must be a high mortality rate.
4. Members of the same species are not identical but show variety in all their characteristics.
5. Because of these variants, some are more successful than others in the competition for survival, and the parents of the next generation will be naturally selected from among these members that show more effective adaptation to the conditions of the environment.
6. Hereditary resemblance between parent and offspring is a fact.
7. Following on from this, subsequent generations will gradually maintain and improve on the degree of adaptation realized by their parents.

The book was a sell out on the first day of publication. Although the religious Victorians were eager to read, they were shocked and disgusted. The religious issue and the implication of man's evolution from ape forms (carefully avoided by Darwin) became greatly disputed and argued.

By 1860 it was time for a public debate. The orthodox followers were represented by the Reverend Samuel Wilberforce, Bishop of Oxford and Darwin by Thomas Huxley, a trained medical surgeon. Although a brilliant man,

Huxley (*left*) and Wilberforce (*below*)

By de-tailing mice for many generations, Weisman hoped to obtain evidence to discredit Darwin.

Wilberforce had no knowledge of the evolutionary issues and cared little for scientific philosophies. Huxley spoke in an effectual way, pointing out the Bishop's petty scientific attitude and explained the new ideas in a precise and factual manner. However, the evolutionary conflict was not completely settled and in certain aspects it still exists tody.

Weisman (1834-1914), a German biologist, was the first major critic of the Darwinian theory of evolution and Lamarckian theory. Between 1868 and 1876 he published a series of papers contending that acquired characteristics of all genetic or body variation could not be inherited. He made little use of practical evidence, although he had carried out a series of crude experiments such as de-tailing mice for many generations to show that a de-tailed mouse never produced a tailless mouse.

He assumed that there was a distinction between a body of an organism – the soma, and the cells concerned only with the reproduction – the germ plasm. He postulated that only the germ plasm could affect inheritance and that the soma could play no part. He also made biologists conscious of chromosomes by identifying them with the hitherto hypothetical particles within them. Weisman's theory was acknowledged and although unacceptable today in the 1880's it meant that Lamarckian views were discredited.

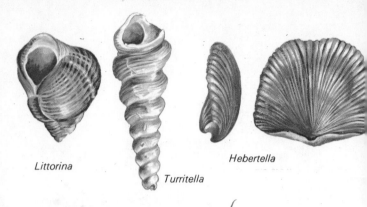

Littorina

Turritella

Hebertella

EVIDENCE OF EVOLUTION ✓

The evidence which helps to confirm the fact of evolution is wide and definite. Much deals with the fossil record of the past, some with present day comparisons of form, and some from actual distribution of animals and plants.

Palaeontology ✓

Unaccountable trillions of animals and plants have lived and died on earth during its long existence. Of these, comparatively few died in a place where their form could be preserved. Nevertheless, there is an abundant wealth of fossil material which has been discovered, although unknown numbers still wait to be unearthed. Fossils are the remains or traces of any recognizable organic structures preserved since prehistoric time.

By far the most abundant are the pollens and spores, and shells of microscopic organisms which settled to the bottom of prehistoric seas. Those with hard parts are more suitable for preservation, and so shellfish such as oysters and lamp shells were often preserved along shore strand lines and in offshore water.

Preservation may take place in several ways. Petrifaction, widely used in reference to fossils, means 'turned to stone', although this very rarely happens to the whole of an animal. Huge fossil logs of the petrified forest of Arizona which became buried in the old Triassic flood plains are good examples. They

Agoniatites   *Stephanoceros*

A selection of fossils

became well fossilized due to the volcanic sedimentary material which covered them as the silica of this material gradually replaced the wood.

Impressions are important fossil forms also. An animal becomes buried in fine-grain sediment and a mould is made by the sediment of the animal's internal or external shape. These internal and external moulds are negatives of the original. The two impressions found in the Solnhofen shales of Bavaria of a Jurassic bird *Archaeopteryx* were formed in this manner. Tracks and paths of animals appear as imprints or compressions. These fossils are frequently clearly displayed on prehistoric mud flats or on moist ground where the animals happened to pass. Dinosaur paths were first discovered in the Connecticut valley, although at first it was thought they were imprints of birds as they were three-toed.

Careful study of rock formation reveals the fact that certain forms are found in certain layers and, with the help of geologists and physicists, acceptable estimates of the age of many deposits of rock containing fossils can be made. Often, a continuous series of fossil forms can be pieced together, as in the horse or elephant, giving details of how the animal changed and developed during the changing periods of the Earth's history.

Fossil evidence is direct in that the fossils are real and depict the exact extinct forms of life in the sequence in which they occurred. A complete sequence tells an undeniable story.

Table showing fossil-bearing geological periods

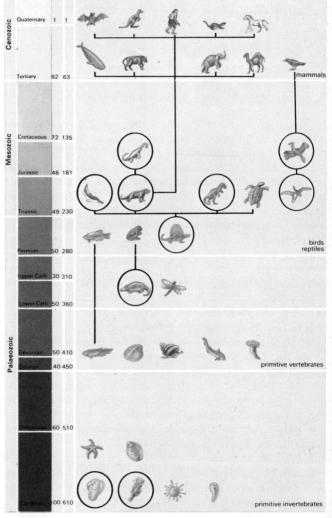

a circle indicates extinction

eras	periods	duration m	years ago m
Cenozoic	Quaternary	1	1
	Tertiary	62	63
Mesozoic	Cretaceous	72	135
	Jurassic	46	181
	Triassic	49	230
	Permian	50	280
	Upper Carb	30	310
	Lower Carb	50	360
Palaeozoic	Devonian	50	410
	Silurian	40	450
	Ordovician	60	510
	Cambrian	100	610

mammals

birds
reptiles

primitive vertebrates

primitive invertebrates

The geological time scale

Using methods based on the radioactive properties of uranium and thorium, modern estimates of the age of the Earth have been given as between three and five thousand million years. Studying fossils has helped to build up and establish a picture of the order in which rocks have been laid down during the course of time, and the history of life. The geological time scale serves as a standard frame of reference for the description and correlation of various isolated events in the history of life on the Earth.

Geological history is broadly subdivided into eras, founded upon the general character of the life they represent. Thus there are the Cryptozoic (hidden life), Palaeozoic (ancient life), Mesozoic (medieval life), and Cenozoic (modern life) eras. Each era is subdivided into a number of periods, based usually upon selected sequences of stratified rocks which display certain fossils. The Tertiary is divided into five periods, according to the relative numbers of fossils representing the more recent life forms that are found in the rocks.

An extremely important fact to remember is that abundant fossil material does not appear until lower Cambrian times. This is a quite late stage of geological time (see pages 72 to 73). The pre-Cambrian era covers over 3,500 million years, representing perhaps nine-tenths of the Earth's history.

In the geological clock the smaller circle represents the age of the Earth (about 5000 m. years) and the larger the last 600 m. years. Each 'hour' represents 50 m. years.

a   mammals

b   reptiles

c   amphibians

d   fishes

e   invertebrates

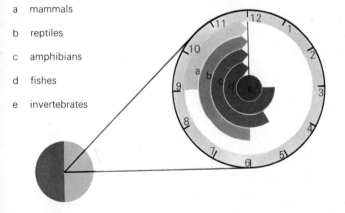

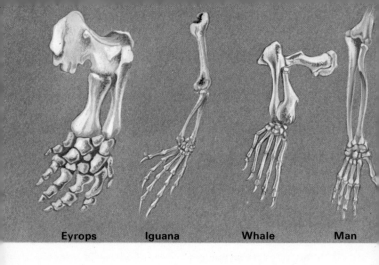

| Eyrops | Iguana | Whale | Man |

Comparative anatomy and vestigial organs ✓

Comparative anatomy reveals the existence of plans of structure in large groups of organisms. However, some similarities in structure do not relate the organisms; this is to say the similar parts are analogous, serving the same purpose. The wings, legs and eyes of a locust are analogous to the same parts of a blackbird in that the two may fly, walk and see respectively with them. On the other hand, similarity by homology, implies structural and developmental likenesses which may not even perform the same function. These similarities are inexplicable unless the animals are descended from a common ancestor.

Structural plans based on homologous relationship nowadays are the basis of the classification of living things. On studying the variety of animal and plant life one finds that each group is characterized by a common form which underlies the group's structure.

Structures have often been discovered which are inexplicable unless one accepts that evolution has taken place. Broadly described as vestigial structures they often appear to be rudimentary or underdeveloped and of no value to the owner. On studying related forms they are clearly homologous with structures most often of great use in other organisms.

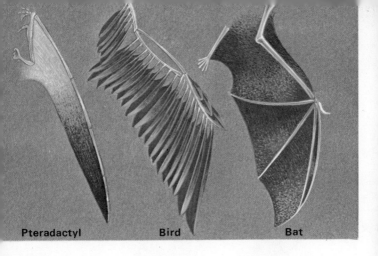

| Pteradactyl | Bird | Bat |

Homologous variation in pentadactyl limbs (*opposite*) and analogous similarities in wing form (*above*)

Snakes, of course, have no visible external signs of legs but internal remnants of hip-girdles and hind limbs surely indicate resemblances to the four-legged reptiles.

Similarly, the whale exhibits well-developed forelimbs in the form of flippers, which can be related to the pentadactyl limb and forelimb of land mammals. However, there are no hind limbs visible, but on dissection a few bones are found embedded in flesh, sometimes in the form of a vestigial pelvic girdle with thigh bones attached, indicating that the whale group evolved from terrestrial mammals.

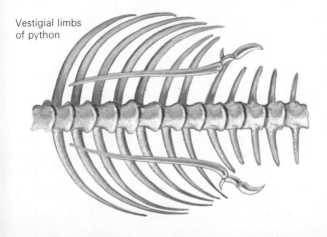

Vestigial limbs
of python

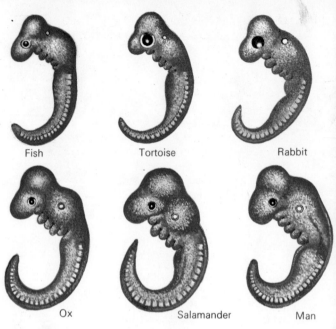

| Fish | Tortoise | Rabbit |
| Ox | Salamander | Man |

Similar embryos of widely differing adult forms

Embryology √

Embryology reveals remarkable similarity of structure between animals in which the adult stages are as different as fish, tortoise, hen, rabbit and man. No one mistakes the adult forms, but it is very difficult to identify embryos without knowing the parents. Von Baer was the first to record observations of similarities but Darwin was the first to point out the reason. If embryos are similar they must have descended from a common ancestor, from which they have inherited the embryonic stages repeated in their own development. It was suggested that during its embryonic development every individual goes through the stages by which it evolved from a primitive form. This is considered today rather an exaggeration but in the embryos of most vertebrates, at comparable stages, they all have fish-like characteristics, with gill pouches or slits in the head. Each develops a vascular system with a single circulation, and the heart is not divided into right and left halves. The gill pouches are more aptly called visceral clefts as they never truly function as gills, but are served in the early stages by arteries undoubtedly homologous with those of an adult fish. As each embryo completes its developmental period, each emerges as an unmistakable adult form.

Intermediates

Comparative anatomy studies have helped to form plant and animal life into related groups and in some cases evidence has been found of certain forms which link one group to another. Some animals cannot be placed definitely in one group as they show characteristics of two groups. They are thus termed intermediates.

The arthropods and the worms can be linked in this manner by the caterpillar-like *Peripatus*. From external appearances it looks half-way between a grub and a worm, the worm-like body bears flexible appendages, but they are not jointed as in the arthropods. Internally it has the arthropod feature of a body cavity containing blood, but it is worm-like in that it has a series of nephridia, and one segment has a pair of legs, similar to the nephridia found in each segment in worms.

Peripatus

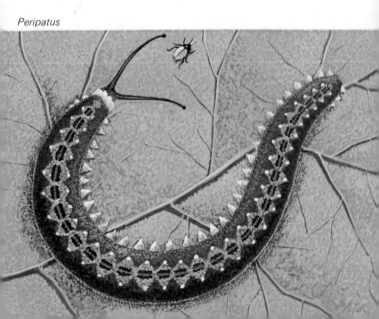

Domestic varieties of dog artificially bred by man all belong to one species.

Immunological evidence

Immunology is the study of animals which are made immune (protected) against a particular infection by immunization. Certain immunological tests help to give evolutionary evidence and relationships between certain animal species. If blood of a species A is injected into another species B, the blood of B acquires the power to precipitate the proteins of the blood of A. It is then found to be able to precipitate the blood of other species related to A, but to a degree which depends on the closeness of the relationship. The degree of precipitation coincides with the degree of relationship suggested from anatomical studies. Immunological evidence is not direct proof of evolutionary relationship, but contributes to the weight of evidence for that view.

Domestic animals and plants

Domestic dogs, cats, horses, cattle and fowls, for example, dramatically illustrate the point that animals can change their form and structure. The collie, greyhound, bulldog, dachshund and poodle would certainly be classed as separate species if found in the wild. Since they freely interbreed if allowed, and as mongrels show all characteristics intermediate between the different breeds, they are actually all one species.

Domestic plants have similarly been bred for their particular characteristics, for example, the numerous varieties of roses. Also, strains of wheat have been developed for their grain yield and rust resistance.

Man has, of course, selected characteristics that are of use to himself, or of ornamental (fancy) design. Nature has selected characters that are useful to the organisms themselves. Therefore, while the two processes are the same in principle, they differ in direction. Owing to this difference, many types of domesticated animals would be eliminated immediately in the wild state.

Classification evidence

Systematics is the study of animal and plant classification. If all species were originally created once and for all distinct, there should be no difficulty in making lists of them. This is not so, however, as species show variation within the kind and if they spread into new environments, they may become adapted and eventually form daughter species. Examples should be visible therefore of evolution in progress now, especially for those species with a wide geographical range. It is difficult to decide between variety and species however. The cat family Felidae, for example, once comprised twenty-five genera whereas today it contains only one or two, depending on authority.

Normal variation within the cat family Felidae

Lion

Tiger

Leopard

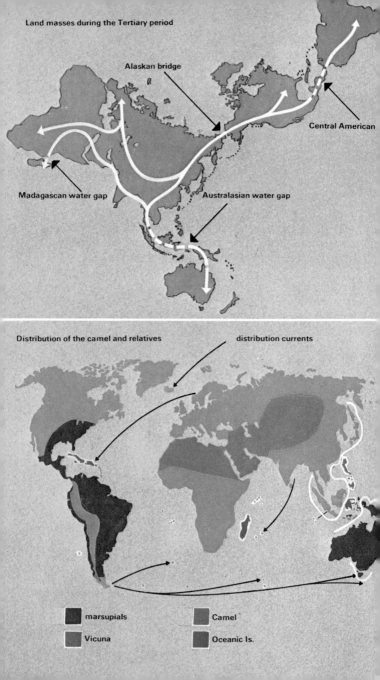

Land masses during the Tertiary period

Alaskan bridge

Central American

Madagascan water gap

Australasian water gap

Distribution of the camel and relatives

distribution currents

marsupials

Vicuna

Camel

Oceanic Is.

Geographical distribution

Evidence of great importance is provided by the facts which emerge from studying the geographical distribution of plants and animals. Regions differ, sometimes very markedly, in their fauna and flora, though they have a comparably similar vegetation and climate. Darwin pieced together information concerning the geographical distribution of various fossil and living animals. He noticed how some related animals were separated by thousands of miles or by a mountain range or some other physical feature. The discontinued distribution of related forms must have risen when animals migrated to their present habitats and became extinct during time in the intervening regions. Fossil evidence has helped to support this view as many fossils have been discovered in parts of the world linking the groups.

Cosmopolitan groups (those spread all over the world with no barriers to check their movement) are small in number. For example, certain insects, bats and birds are widespread in distribution as they can fly. However, most species today have a definite area of distribution, due to the formation of barriers set up as the Earth's surface and climate changed.

Although the main land-masses varied little during the Tertiary period, there were considerable changes in communication between them, both by the making and breaking of narrow land bridges and by sharp temperature gradients and desert areas. The various regions are marked on the map and the land bridges that allowed animal migration.

The Ethiopian region was separated from the Palaearctic region by the sudden change in temperature and desert conditions in North Africa. The isthmus between North and South America was broken in the Eocene period until the Pleistocene. Thus the South American fauna shows certain distinctions from other regions. Australia, east of Wallace's line, has been cut off since the late Cretaceous.

There is evidence that the many forms of animal life evolved in the central Holarctic areas and migrated away towards the extremities. Several of the types which evolved quite early remain as vestiges at the outer areas of the southern land masses, such as the marsupials, monotremes and lungfish. The distribution of the camel, vicuna and marsupials is shown.

MECHANISM OF EVOLUTION

Mendel

The name 'Mendel' is usually associated with 'genetics'. Who was this Mendel and how did he fit into the evolutionary scene? Mendel made his discoveries around 1860, although his work remained unnoticed until the turn of the century, nearly twenty years after his death.

Johann Mendel was born in 1822 in the Silesian village of Heizendorf where his parents were peasant farmers. His early schooling and university course showed him to be a brilliant young man and he became devoted to his philosophy studies at Olmütz University. Due to lack of food because of money shortage and overwork Mendel suffered two breakdowns in health. It became agreed in 1843 between Johann Mendel, his family and his Professor of Physics, Friedrick Franz, that he should become a novice in the monastery in Olmütz. On October 9th, 1843, he was admitted under the name of 'Gregor'.

Life was still difficult and Gregor Mendel had many health and mental troubles during his monastic life. He became a lay brother and then a fully qualified cleric in the town of Brünn. He also distinguished himself teaching in the local school. From October 1851 to August 1853 under his Professor's direction he once more became a student, this time at Vienna

De Vries, with others, later acclaimed Mendel's work.

Mendel (*opposite and below*) From the results of his plant breeding experiments Mendel proposed two laws of genetics

University, with the purpose of obtaining a more thorough grounding in the discipline of natural science.

He returned to teaching but also began to experiment with plant breeding. From thirty-four varieties of edible peas he selected twenty-two for experiments to determine the statistical relations of the various hybrid offspring. Mendel's whole theory simply predicted the number of different forms that would result from the random fertilization of two kinds of egg cells by two kinds of pollen grains.

Having satisfactorily concluded his experiments, Mendel put his findings into two papers entitled 'Experiments on Plant Hybridization'. No stirs were caused when he read these to the Brünn Natural Science Society in 1856. When he was appointed Abbot of the Monastery in 1868 he found little leisure time to continue his research, and in January 1884, he died.

His work lay neglected until 1900 when three independent investigators, De Vries in Holland, Correus in Germany, and Tscherenmak in Austria, found Mendel's forgotten paper and proclaimed its importance.

Mendel's first law

Mendel set about investigating the problem of the manner in which free-breeding variations within a species are related to one another. In a number of garden plants he noticed that within each species certain definite variations could be observed. In the garden pea, *Pisum sativum*, he saw many characteristics: for example, there were tall (6 to 7 feet) and dwarf (9 to 18 inches) varieties; their seeds could be smooth and rounded or wrinkled in appearance; the unripe pods could be green or yellow. His approach of basic biometric statistics fitted in with his physics background and the results had to be analysed and mathematically interpreted. Mendel knew the life histories of the plants and thus crossed in one experiment a dwarf variety about a foot high with a tall variety reaching about 6 feet. He obtained hybrids that were all tall. He pollinated the hybrids with their own pollen and obtained 75 per cent tall and 25 per cent dwarf, that is a ratio of three to one. When he self-pollinated this second generation he found all the dwarf variety bred true in the following generations. However only one-third of the tall variety bred true, the rest gave 75 per cent tall and 25 per cent dwarf again, and so the process continued.

From many experiments using different characters Mendel realized there must be some characters (such as wrinkled seeds or dwarfness) which can, during mating, be 'masked' or overcome when in the company of certain other characters (such as round seeds or tallness). Although 'masked' in some given generation, they retain the power to reappear in a later generation. For these characters he used the term 'recessive'. Those characters that could mask the recessive characters he called 'dominant'. If the two were coupled together the dominant character only would be visible and has the effect of preventing the recessive displaying its effect or form.

Thus, from the third generation results Mendel worked out that one third of the tall plants had no dwarf factors being masked and therefore bred true. The others would be masking the dwarf recessive factor which when the plant was self-fertilized would give the three to one ratio again.

The seed counts and interpretation enabled Mendel to propose what is known as his first law which states that *characters*

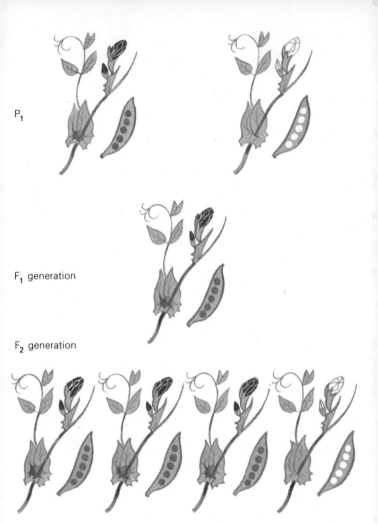

P₁

F₁ generation

F₂ generation

In this cross, a character for seed form is suppressed in the F₂.
When plants of this generation are self-pollinated however, the
character reappears in a quarter of the next generation.

*are controlled by pairs of factors which do not blend during life
and which pass into separate cells during reproductive processes,
prior to fertilization.*

Mendel's second law

Mendel did not end with investigating just one pair of contrasting characters, but went further, and noted the results when plants with two pairs of factors were self-fertilized.

One must remember that although Mendel was not familiar with chromosomes, he saw the need for finding out how characters would behave in relation to each other in their passage from generation to generation.

In one study, he chose seed form and seed colour as his subjects, using a pure breeding plant with round and yellow seeds as one parent, and a pure breeding plant with wrinkled and green seeds as the other. His method was the same as that previously described, but he recorded the frequencies with which each character appeared with the other, instead of just separate frequencies. His results showed that in the first filial generation all the plants were round and yellow seeds, indicating the dominance of round over wrinkled, and yellow over green characters, respectively. After self-fertilization of these seeded plants, Mendel found he had a second filial generation showing four different combined sets of characters.

He calculated the proportions and found them approximately in the ratio of nine round and yellow seeded plants, three round and green seeded plants, three wrinkled and yellow seeded plants and one wrinkled and green seeded plant.

From these results, Mendel deduced a second law of inheritance. He stated that *each of a pair of contrasted characters may be combined with either of another pair during the process of reproduction so that there is complete independence of combination among the factors present.*

In all, Mendel studied seven pairs of characters in peas, involving seed colour, seed surface, flower colour, vine height, colour of unripe pods, pod shape and position of flowers. These crosses which involved two character differences separable in inheritance are called di-hybrid crosses (a cross involving a single pair of alleles is monohybrid). Similar experiments of di-hybrid inheritance have been carried out in *Drosophila*, the fruit fly, with the same ratio results. The particular combination in which characters are brought into a cross makes no difference at all to the manner in which they

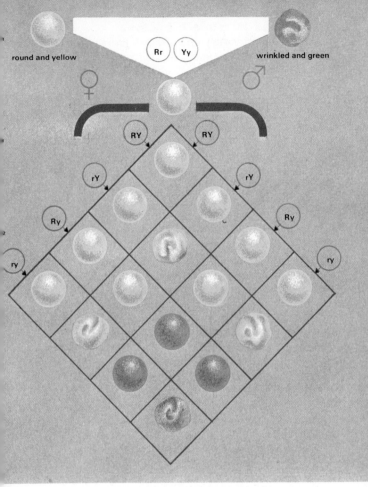

round and yellow

Rr Yy

wrinkled and green

RY   RY

rY   rY

Ry   Ry

ry   ry

Results of a di-hybrid cross involving seed form and colour

are assorted and recombined in the F2. If both dominant characters are brought in by one parent and both recessives by the other, the ratio results are the same as when both parents have one dominant and one recessive. The results are always 9/16 dominant characters, 3/16 one dominant one recessive, 3/16 other dominant other recessive, 1/16 recessive.

Terms used in modern genetics

Alleles (allelomorphs) Characters or genes which follow the Mendelian laws. Allelomorphic characters are those such as smooth or wrinkled seeds, or tallness and dwarfness of stem, which can be paired up from the point of view of inheritance. Allelomorphic genes are the representatives in the gametes of these characters, producing their different effects on the same development process.

Back cross The crossing of a heterozygous individual of the F_1 back to either of the homozygous parents.

Crossing-over Interchange of corresponding chromosome sections by homologous pairs of chromosomes giving recombination of linked genes.

Di-hybrid ratio Two pairs of contrasting characters are crossed. Mendel's F_2 proportions 9:3:3:1 represent the di-hybrid ratio in which there are sixteen possibilities.

Diploid A full set of paired chromosomes (twice the haploid number); characteristic of normal cells.

Dominant A gene which produces the same character when it is present in a single dose together with a certain allelomorph (recessive gene in heterozygous condition), as it does in double dose (homozygous), is said to be dominant over that allelomorph.

F_1 generation (first filial generation) The offspring resulting from crossing the plants or animals of parental generation.

F_2 generation (second filial generation) The offspring resulting from crossing members of the F_1 among themselves.

Gametes Sex cells carrying the haploid number of chromosomes.

Gene A unit of inherited material, first called 'gen' by Johannsen in 1909. It corresponds to Mendel's germinal unit or factor which he supposed was the cause present in the gamete for the appearance of a character in the adult.

Haploid A single set of unpaired chromosomes, characteristic of gametes.

Heterozygous, homozygous Terms applied to the genetic constitution of an individual with respect to the possession of a particular pair of allelomorphic genes. Mendel's original parent plants are said to be homozygous, those of the F_2 generation which possessed two factors for tallness, or two factors for dwarfness are also homozygous, while those of the F_1 generation possessing one factor for tallness and one for dwarfness are said to be heterozygous.

Hybrid The result of a cross between parents showing unlike characters. Mendel's F_1 generations were hybrids and did not breed true.

Independent assortment The chance distribution to the gametes of allelomorphs. The distribution of members of one pair having no influence on distribution of members of another.

Linkage Association of two or more non-allelomorphic genes, so that they tend to be passed from generation to generation as an inseparable unit, and fail to show independent assortment. This is due to the fact that they are on the same chromosome.

Meiosis Division of sex cells forming gametes during which the chromosome number is halved.

Mitosis Division of cells without change in chromosome number.

Monohybrid ratio The ratio between the numbers of individuals possessing different genetic constitutions in the F_2 as a result of crossing a single pair of contrasting characters. Mendel's F_2 proportions, one homozygous tall: two heterozygous tall: one homozygous dwarf, represents the monohybrid ratio in which there are always four possibilities.

Phenotype Observable characters shown on an organism irrespective of genetic constitution, for example tallness or dwarfness.

Pure line A succession of generations of organisms homozygous for all genes. Continued self-fertilization will rapidly lead to a homozygous condition in all species, since all factors already homozygous cannot change. Of the heterozygous allelomorphs, half become homozygous at each generation. Thus, the number is steadily reduced.

Reciprocal cross Crosses between parents in both directions. The whole of Mendel's results were based on such crosses.

Recessive The converse of dominant. A recessive gene has no effect on the phenotype unless homozygous. For example tallness is dominant over dwarfness which only shows if in homozygous conditions.

Zygote Cell formed by the union of two gametes.

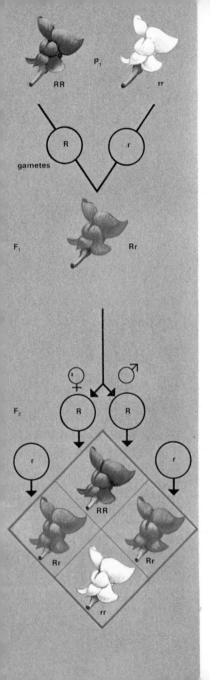

Incomplete dominance

Not all characters behave as perfectly as Mendel's laws suppose.

Incomplete dominance indicates an apparent failure of one allelomorphic character to dominate the other in the F_1 generation. In the Andalusian domestic fowl, it is possible to obtain true breeding strains of the black and white forms. When crossed, the first generation offspring are neither black nor white, but an intermediate colour called 'blue' or 'splashed white'. If these fowls are allowed to breed among themselves the next F_2 generation shows a segregation into blacks, blues and whites in the proportions $1:2:1$. It is now known that if a gene for black colour is present with a gene for white colour, then a single black gene is unable to produce full black pigment development. A fowl must possess genes from both parents for full black pigment production.

In plants the cross between red and white Snapdragon (*Antirrhinum*) gives hybrids of an intermediate pink colour. This again shows that when the alleles R and r come together in a cross neither is dominant, and the heterozygote Rr is pink flowered.

P₁

blue or splashed white

F₁

F₂

F₂

F₃

F₃

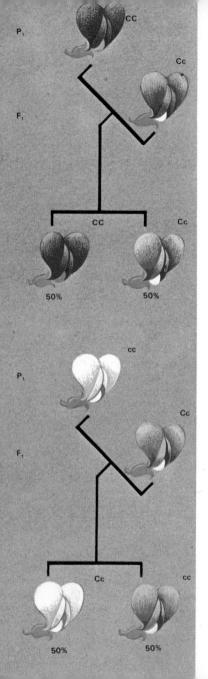

Back crosses

Mendel also examined allele behaviour in back crosses. A heterozygous F_1 individual is crossed back to either of the homozygous parents. The red-flowered F_1 hybrid from the cross of red and white peas is crossed back with the dominant red-flowered parental variety in one, and the recessive white-flowered parental variety in the other. The red-flowered F_1 hybrid, according to Mendel, is a heterozygote Cc and the white-flowered parent a homozygote cc. The gametes produced by a hybrid heterozygote are always pure. Therefore, in a red-flowered F_1 plant one half should carry C and the other c. The gametes of a white-flowered plant which are recessive are all c.

The progeny in the back cross to the red-flowered parent are all red-flowered plants and if Mendel's theory is correct they should consist of 50 per cent homozygous red-flowered (CC) and 50 per cent heterozygous red-flowered (Cc). This is verified by looking at the back cross to the recessive white-flowered parent. White-flowered (cc) and heterozygous red-flowered (Cc) are produced in equal numbers.

Lethal crosses

According to Mendelian segregation, gametes that carry different alleles of the same gene unite at random and give a ratio of three dominants to one recessive. However, these conditions are not always realized.

Matings between mice with yellowish fur have produced offspring with yellow and non-yellow fur in a ratio of two to one. Furthermore, matings between yellows give litters which are smaller in number by about one quarter than litters from yellow and non-yellow parents. Apparently zygotes homozygous for yellow are inviable and this hypothesis has been verified by scientists crossing yellow females with yellow males. They found that some embryos die in an early stage of development. From the illustration:-

  Ay = a dominant gene for yellow fur
  a = its recessive allele for non-yellow

The allele Ay is lethal when homozygous but when present with the non-yellow recessive allele a, it merely modifies the colour of the fur to a yellowish tinge.

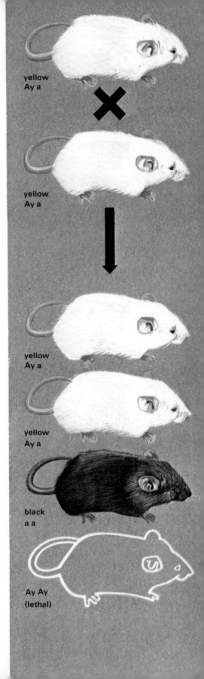

yellow
Ay a

×

yellow
Ay a

yellow
Ay a

yellow
Ay a

black
a a

Ay Ay
(lethal)

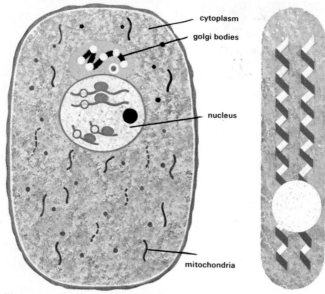

Diagrammatic cell and chromosome

Cells and chromosomes

The living substance of both plants and animals is organized into microscopic units called cells. Cells were first seen by Robert Hooke as early as 1665, when he observed and described the cells of a piece of cork.

Usually the cell consists of a dense body, the nucleus, enclosed in its own membrane and surrounded by less dense colloidal fluid the *cytoplasm*. The cytoplasm contains a variety of structures including the *mitochondia,* centres of enzyme activity in oxidative metabolism, only just visible through a light microscope. Animal cells contain *golgi bodies*, function uncertain, and *centrosomes,* important in cell division. In the nucleus is found the structures of great genetic importance, the *chromosomes*.

Certain cells of the body carry out division for either growth by *mitosis,* or for the formation of sex cells by *meiosis.*

Mitosis is the process by which a cell divides into two during growth during which the chromosomes undergo definite changes. These are the recognized stages:-

Interkinesis The cell is not dividing and separate chromosomes are not usually distinguishable.

Prophase The cell is preparing to divide and the chromosomes become clearly visible as threads which gradually shorten and thicken by coiling.

Metaphase The nuclear membrane disappears and the spindle appears. Arranged around the central plane of the cell, each chromosome splits longitudinally. Each cell has, in effect, set of 'double' chromosomes.

Anaphase The centromeres split and the chromatids move towards the poles.

Telophase The chromosome halves move to the poles forming one complete set at each end.

The cell now finally divides, the spindle disappears and the chromosomes lengthen. 'Daughter' nuclei form, each containing a new set of chromosomes and the conditions of interkinesis are restored.

Stages of mitosis illustrated diagrammatically

prophase

metaphase

metaphase

anaphase

telophase

interkinesis

Stages of meiosis

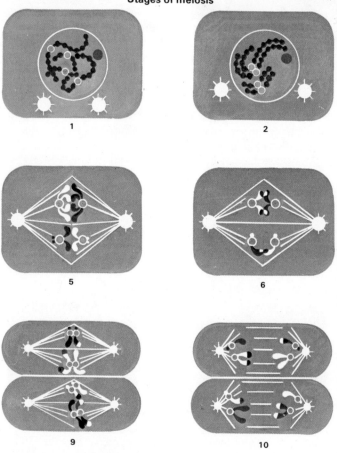

Meiosis is not concerned with growth. It is a known fact that the chromosome number of a species is always fixed. How is it that the union of the parent cells does not produce a doubling of the chromosome number? At some point during sexual reproduction, the chromosome number in the sex cells has to be halved and this happens during the formation of the male sex cells *spermatozoa*, and the female sex cells *ova*, by a special method of cell division called meiosis.

Meiosis differs from mitosis in that it produces four new

3

4

7

8

11

12

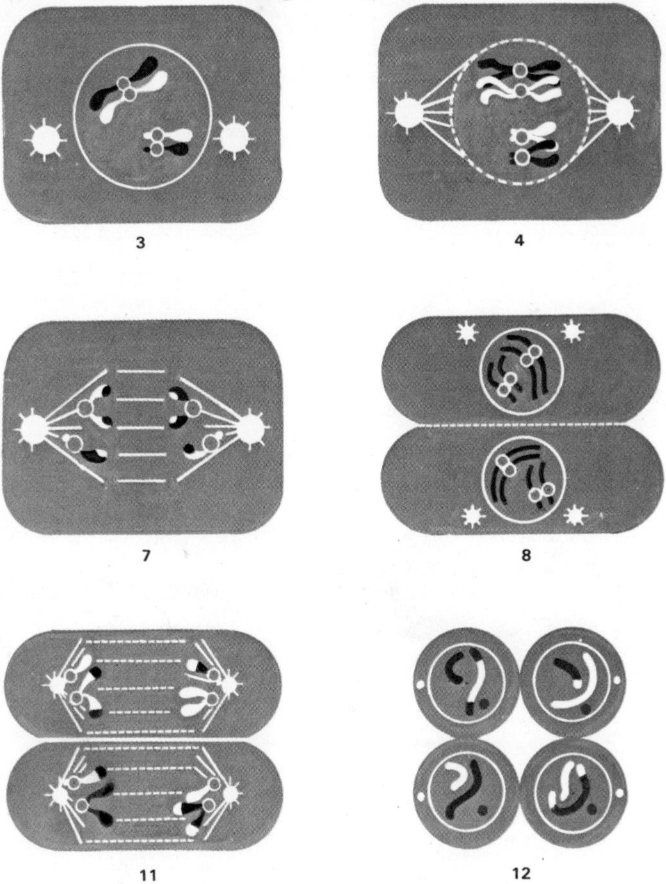

cells in each of which the diploid number is halved. The process is brought about by two nuclear divisions, but only one duplication of the chromosomes. We refer to these two divisions as the first and second meiotic divisions. During the first meiotic division, the phases occur in the same sequence as they do in mitosis. However in the prophase of meiosis, there is a significant difference in the behaviour of the chromosomes. A number of distinct stages within the first meiotic division can be recognized.

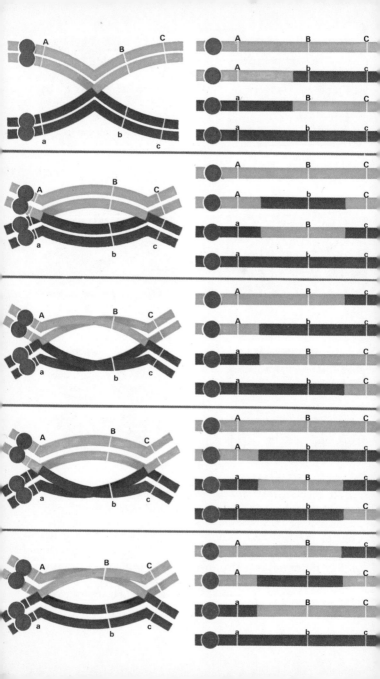

Linkage and crossing over

It is now known that the chromosomes carry the genes responsible for the inheritable characteristics. As the fixed number of chromosomes for each species is quite small, each chromosome must carry a fantastic number of genes. The genes are arranged in linear order, along the length of the chromosomes and in 1910 Morgan discovered many facts about their positions. He supposed the tendency for linked genes to remain in their original combinations was due to their residence on the same chromosomes. He also thought the degree or strength of linkage depends upon the distance between the linked genes in the chromosome. This basic idea has now led to the construction of genetic or linkage maps of chromosomes showing the exact position on which chromosome an inheritable characteristic is found.

If the genes are always in the same place then two genes situated on the same chromosome should remain together in all cases. However, although genes are bound together, the chromosomes when pairing during meiosis can exchange genes by 'crossing-over'. This interchange of sections of chromosomes leads to the recombination of linked genes and so some alleles switch from one chromosome to another. The chromosomes pair with remarkable preciseness during the prophase of meiosis and this is evidently brought about by mutual attraction of the parts of the chromosomes that are similar because they contain allelic genes. Between the **pachytene** and **diplotene** stages the paired chromosomes each divide into two chromatids so that there are four in number.

At the first division of the chromosomes, the newly formed chromatids exchange one or more links or *chiasmata*. At each chiasma, two of the four chromatids become broken and then rejoined so that the new chromatids are compounded from sections of the original ones. The new chromosomes that arise as a result of meiosis carry genes that before meiosis were located in different members of the pair of chromosomes. However, crossing-over does not occur in the majority of gametes. Usually, about 97 per cent of the gametes contain the parental combination, and 3 per cent contain recombinations.

Chromosome cross-overs and resulting gene recombinations

Determination of sex

How is it every child has a mother and father, but some children are males and others are females? What determines whether an offspring will be a male or a female? Questions like these puzzled non-biologists and biologists alike for centuries and before 1900 most theories were just wild guesses.

Heredity was thought to determine merely the similarities between parents and offspring, and the heredity of a child was supposed to be a compromise between the inheritable characteristics of the parents. How did sex fit into such ideas of heredity? Could there be any relation or control between the two? The vital evidence necessary to answer the sex determination question was not unfolded until the early 1900's by genetic researchers.

It was observed in some insects that male chromosomes gave an odd total number, and that females had an even number. Another observation was that in some males there was often one pair of chromosomes made up of unequal

Unequal chromosome pair (*left*) and normal pair (*right*)

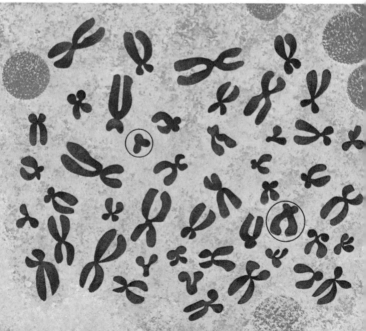

partners, whereas in the female this pair was formed by equal chromosomes. The unequal chromosomes are called the *XY* pair and the matching chromosomes the *XX* pair. When meiosis is completed in the male, there are two kinds of spermatozoa, one with *X* chromosomes, the other with *Y* chromosomes. In the female only a single egg with an *X* chromosome is present. The sex of an individual is, accordingly, determined at fertilization. If an *X*-carrying spermatozoon fertilizes the female egg, the union will produce a *zygote* with two *X* chromosomes, developing into a female. A *Y*-carrying spermatozoon gives an *XY* zygote which becomes a male. The chromosomes obviously differ in the male and female and so it follows that the genes will differ.

In man, other mammals, and most insects, the heterozygous sex is the male. However, although the principle is much the same, the chromosome mechanism of sex determination has been found to vary in different organisms. In birds, butterflies, moths and some reptiles and amphibians, the *XY* zygote produces a female and an *XX* zygote produces a male.

Mechanism of sex determination in human offspring

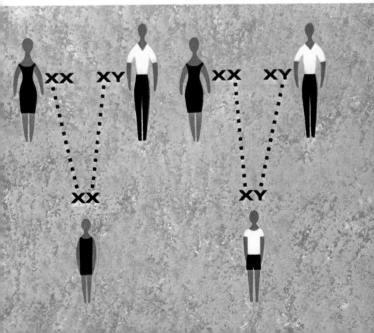

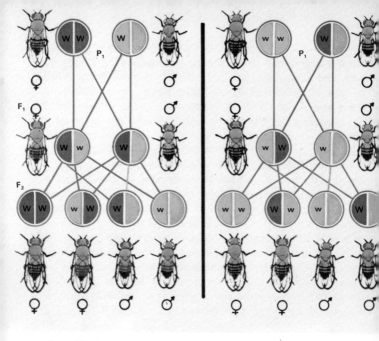

Sex linkage

The sex chromosomes, it was found, did not just contain genes which controlled the sex of the individual. They had many genes concerned with other characters of the individual. Morgan was the first to discover this from his work with the fruit fly, *Drosophila*.

Morgan found one individual breeding with normal red-eyed wild *Drosophila*, which had white eyes. The white-eyed *Drosophila* bred true in subsequent generations, and a new variety was established. He crossed this new white-eyed variety with the wild red-eyed type. The results from a cross of a white male and a red female were different from those obtained from the reciprocal cross of a red male with a white female.

From the cross of white-eyed male with red female, the first generation all have red eyes. When these are bred together, one quarter of the offspring have white eyes, indicating that red-eyed allele is dominant over white-eyed. But of the F₂ offspring half of the males only are white-eyed, the others red-eyed but all the females are red-eyed. Although the females do not show any differences externally, their sex chromosomes contain two different kinds of genes. This is shown when the two types of females are bred as all the off-spring are red-eyed, except for about one eighth of the total which are white-eyed males. This results from the half of the females carrying the recessive white-eyed gene which appears when paired with a male sex chromosome.

Different results, of course, can be expected when a red male is bred to a white female. By studying the illustration it can be seen that the F_1 offspring consists of the red-eyed females and white-eyed males. These produce F_2 offspring of red-eyed and white-eyed flies in equal numbers in both sexes. All the white-eyed flies breed true, as do the red-eyed males when bred to pure red-eyed females. But red-eyed F_1 females must be heterozygous for when bred to either white or red males, half the male offspring are white-eyed.

Studying the *Drosophila* crosses it can be seen that the sex-linked trait of white eye colour follows a criss-cross inheritance. A male transmits his sex-linked trait to only his daughters where it is recessive.

Reciprocal crosses to show sex-linked inheritance of eye colour in *Drosophila* (*left*) and colour blindness in man (*below*)

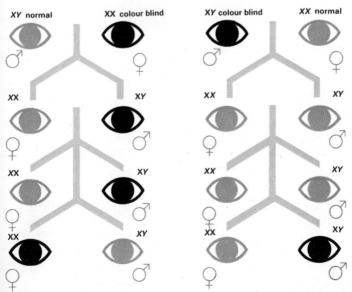

Altitude induced variation in *Horkelia california*

Variation

Slight variations can be caused by crossing-over during meiosis (see pages 48 to 49). This has helped during the course of evolution to provide different characters in a species. Due to the forces of natural selection if certain groups of a population become isolated by geographical, climatic or other factors, a gradual divergence occurs until differences between the populations prevent them from interbreeding. They have thus diverged to become separate species. Inherited variation acted upon by natural selection and aided by environmental isolation can therefore be said to cause evolution. Even with individuals of the same species slight variations are usually visible. Variation may be caused by changes in the hereditary material, but also by changes in the environment. Environmental factors such as soil, climate, food supply, humidity and altitude may produce changes known as modifications which are not inherited however, and play no part in evolution.

Transplanting experiments often illustrate the effect of the environment on a plant's growth. A single plant is divided into pieces so that a number of plants of the same age and genetic constitution are obtained, forming a *clone*. The individuals of the clone may be grown in different climatic conditions to study the effects of the environment.

Transplanting experiments of *Horkelia californica* to different altitudes produce interesting results. This plant which grows wild at 200 metres in California, reaches a good height at sea level, but is remarkably stunted in its growth at higher altitudes, often perishing during the winter months.

The amount of food an animal or plant consumes or builds up is another extremely important environmental factor which helps to produce variation. Pigs are illustrated below as an example. Those able to eat as much as possible get very large and fat, while those put on a diet are long and lean. This experiment was performed in a pure strain of pigs where hereditary variation can be discounted. Some were given restricted amounts and others unrestricted amounts of the same food for equal periods of time. Differences were found in the development of bones, those on the unrestricted diet at sixteen weeks being much larger with massive bones. When full grown it was discovered that those on a restricted diet had longer and narrower skulls. They also had a longer lumbar region of the vertebral column (loin) and larger hip girdles than the unrestricted pigs, although taking longer to become fully developed.

Variation in size of pigs fed on unrestricted diet (*top*) and restricted diet (*bottom*)

Geographical isolation

The birds of the Galapagos Islands illustrate a remarkable example of evolution and adaptive radiation. The life to be found on the Islands gave Darwin some of his first facts in his study of evolution. The animal and plant life probably found its way by chance transportation across the ocean. It includes several species of birds, a rat, a bat, a giant tortoise, iguanas, a snake, one lizard and one gecko, no amphibians, and a limited number of land insects and snails.

Small numbers meant that competition was less and the species would occupy not only their usual habitats, but in addition, others which were not filled with rivals as in their mainland environment.

Studying the life of the thirteen larger islands, which are separated by up to 100 miles, it can be seen that several of the Galapagos animals have formed separate island races. The groups of fourteen species of finches illustrates extreme island differentiation. They must have been in the archipelago for a considerable time forming quite distinct races, and radiating to form a series of birds occupying quite varied habitats often quite unfinch-like in behaviour and form.

The beaks of the finches vary according to their diet. The five ground finches closely resemble the South American finch species which feeds mainly on seeds. Two of these have left the ground to feed on the prickly pear cactus, *Opuntia*. Those living in trees have beaks adapted for insects although one is a vegetarian. Among this group is a woodpecker-like finch, *Camarhynchus pallidus*, which climbs trees woodpecker-fashion and digs insects from the bark with a stick. The warbler finch group, Certhidea, have long slender beaks for eating small soft insects and have behaviour habits similar to the true warbler. They have evolved to fill this role because until recently there had been no warbler species on the islands. The fourth group has only one member, the Cocus finch, found only on Cocus Island. Various subspecies and races are recognised throughout the archipelago, differing only in minor features of colour or size.

The groups of finches (*top*) observed by Darwin in the Galapagos Islands (*opposite*)

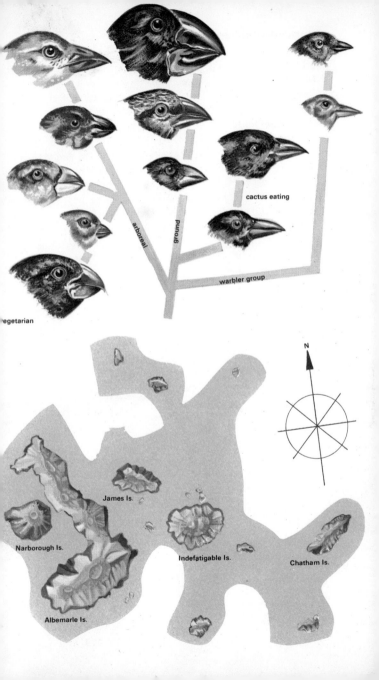

cactus eating

arboreal

ground

warbler group

vegetarian

James Is.

Narborough Is.

Indefatigable Is.

Chatham Is.

Albemarle Is.

N

Geographical variation

When a species covers a large area, on close observation it is found that it gradually changes in certain characters over different parts of the range. The change may be in the size, colour, shape or other features. Such gradients of characters are called *clines*, a term introduced by Sir Julian Huxley, a well-known contemporary biologist and evolutionist.

The Lesser Black-backed Gull and the Herring Gull occupy a ring-shaped range around the North Pole. On the shores of the Arctic Ocean in Eastern Siberia lives the Vega Gull (*Larus argentatus vegae*) with dull flesh-coloured legs and a dark mantle. Eastwards across the Bering straits to North America, the Vega Gull grades into the American Gull (*L. a. smithsonianus*) and across the Atlantic into the British Herring Gull (*L. a. argentatus*) with flesh-coloured legs and a light mantle.

In a westerly direction from Siberia, the Vega Gull grades into the Lesser Black-backed Gull (*Larus fuscus fuscus*) and its

Gulls showing graded characters over their range

Length of wing (mm)

158-177

155-166

135-145

Size cline of the Puffin

British race (*L. f. graellsii*). Thus, there is a continuous gene-pool of gulls around the North Pole, with the gulls interbreeding around the chain. However, the ends overlap in Britain with the Lesser Black-backed Gull and the Herring Gull living together, but differing in behaviour and appearance. The Lesser Black-backed Gull breeds inland and is migratory in winter, the Herring Gull nests on cliffs and is resident. They do not interbreed, behaving as distinct species.

The Puffin provides a good example of a size cline. From the Balearic Islands in the Mediterranean to Spitzbergen in the Arctic, the Puffin varies in a gradual change. The maximum wingspan of these birds increases at the rate of 1 per cent of linear measurement per 2° latitude north.

This is also an example of Bergman's Rule according to which warmblooded animals increase their size in colder climates. This is an adaptation to reduce heat-loss, since the larger the animal, the smaller is the ratio between its volume (which produces heat) and its surface (which loses heat).

	Rana ridibunda (western)
	Rana ridibunda (eastern)
	Rana esculenta

Ecological isolation

Biological isolation is a necessary condition if species are to diverge and new species form. As we have seen, it may be geographical when physical features such as mountain ranges separate whole populations. Also extremely important is ecological isolation, where different modes of life such as feeding, breeding, and nesting habits, or adaptations to different temperatures or degrees of salinity, separate portions of populations.

Rana esculenta and *Rana ridibunda*, species of edible frogs, provide an example of overlapping geographical distribution. The range of the latter overlaps the former on both sides with an eastern race and a western race. The species do not normally interbreed, as there is a reproductive isolation barrier – *Rana ridibunda* breeds earlier than *Rana esculenta*.

Genetic isolation

Related plants occurring side by side may be isolated by their inability to breed and produce fertile offspring. Incompatibility between the gene and chromosome mechanisms of the two plants is often the cause of this.

When *Primula verticullata* is crossed with *Primula floribunda* hybrid offspring are produced, but they are sterile because the chromosomes of one parent species are incompatible with those of the other. The matching of the male set of chromosomes with the female set is difficult in these conditions and the formation of the sex cells is thrown out of gear. Occasionally, however, the hybrid plant undergoes what is known as *polyploidy*. This is a doubling of the chromosomes, and when it occurs the hybrids are able to breed with other similarly formed hybrids, but are sterile with both parent species. The hybrid then breeds true, with a different structure and habit from each of the parent species. It thus fulfills all the criteria of a new species and is called *Primula kewensis*. Similarly the Paeony can be found in the diploid, triploid, and tetraploid form.

Overlapping ranges of edible frogs (*opposite*) and polyploidy in Paeony (*below*)

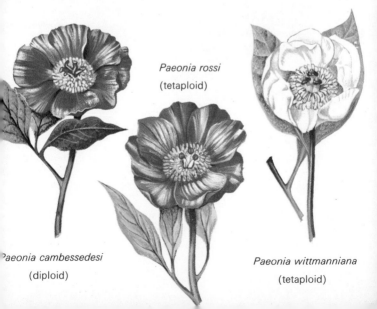

Paeonia rossi
(tetaploid)

Paeonia cambessedesi
(diploid)

Paeonia wittmanniana
(tetaploid)

Male lyre-bird

Sexual selection

In the breeding season of many insects, fish, birds, and mammals, the males show ornamental features and displays. In only a few instances are they borne in the females. Darwin explained the varying appearances in his subsidiary theory of sexual selection. He argued that before mating occurs, the female has the choice of many suitors. She chooses one or more and each chosen will show slight differences from the rest. These characters of the fortunate chosen males will be inherited and appear in the male offspring, as the genes will be on the sex chromosomes and thus sex-linked. Darwin's argument and explanation is that these individuals with well developed striking colours, structures or elaborate behaviour benefit from greater breeding success and more offspring. There are many objections to Darwin's theory. In the majority of cases these secondary sex characters are not the result of sexual selection in Darwin's sense, so much as of natural selection favouring general reproductive activity through conspicuous recognition marks, virile behaviour and fighting ability, etc.

This phase of selection does not work by the complete elimination of the least fit, but by eliminating reproduction of the least fit. It is most effective among the males, as the females usually manage to be fertilized, but there may be a great battle among the males for breeding territories and the possession of the females.

The male lyre-bird shows his display material in remarkable specialized tail feathers. He spreads his tail forwards over his head and vibrates the feathers until they resemble a shimmering haze of light.

In spring, the male stickleback develops his courting colours of a bluish-white back and a brilliant red throat and belly. He stakes out his home territory, builds a nest, defending his rights from all other males and then tries to entice an egg-bearing female by a most elaborate zig-zag dance. If successful the male terminates his courting behaviour by fertilizing the eggs laid by the female.

Male stickleback enticing female to nest

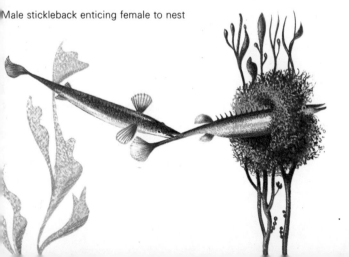

Adaptation

Adaptation displays features in an animal or plant which enable it to survive in the environment in which it lives.

The woodpecker, for example, has four main adapting features. The four well-clawed toes are formed so two face forwards and two backwards, so they can firmly anchor themselves to the tree; the tail feathers are stiff and can be used to help prop the bird securely against the tree; the beak is long and strong so that holes can be drilled in the bark; and the tongue is very long, enabling the bird to reach and catch grubs at the bottom of the holes.

Many other creatures have evolved adaptations for flight, for example, insects, birds and bats. In each case of adaptation the wings are of different construction, but all are successful in flight.

Many species of desert plants, although not closely related, are quite similar in general appearance. This is because they are similarly adapted to keep loss of water by evaporation to a minimum. They all have a tough epidermis, reduced leaves, thickened stems and smooth surfaces as protection against desiccation and death.

Woodpecker (*above*) and desert cacti (*below*)

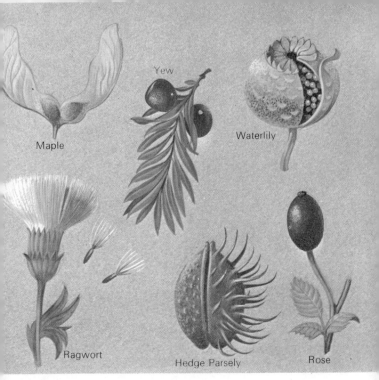

Adaptations for seed dispersal in plants

The survival of a plant species depends to a large extent on the reproductive process and the efficient dispersal of the seed. It is advantageous to the species that the seeds are carried some way from the parent plant, to obviate competition for food, light, and water. Many natural agents do this, such as wind, water, birds and animals.

Some seeds are so light that they are easily blown by the wind, for example those of the Ragwort; other heavier seeds have structures enabling them to fly, such as the Maple.

Dispersal by water is uncommon; however in some plants, for example the lily, the seeds have a spongy covering enabling them to float.

Many seeds have tiny hooks so that they adhere to passing animals and are rubbed off far from the parent plant. Others are brightly coloured such as the rose hip, to attract birds.

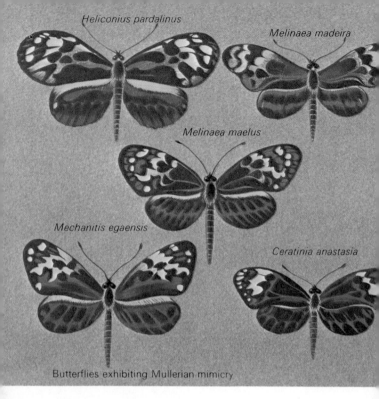

Heliconius pardalinus

Melinaea madeira

Melinaea maelus

Mechanitis egaensis

Ceratinia anastasia

Butterflies exhibiting Mullerian mimicry

Mullerian mimicry

Mullerian mimicry is the term applied to the resemblance found between different species that are distasteful to their potential predators. For example, several species of butterflies resemble each other sufficiently closely to be indistinguishable to birds. Thus, the predator bird, after one distasteful experience, learns to avoid this colour scheme with the result that all the species benefit and the overall losses are greatly reduced. The more species in a 'mimicry ring' the greater the advantage and less eaten, but some butterflies must be sacrificed in educating each generation of birds. The resemblance between the forms is not necessarily very exact, as it is in Batesian mimicry, as the colour patterns are the reminders. A mimicry ring of South African butterflies is shown above. Mimicry is more common in the tropics than in temperate regions.

Papilio dardanus (meriones male)   *Papilio dardanus* (meriones female)

Hippocoon   Niobe   Cerea

Amauris niavius   *Amauris echeria*   *Bematistes tellus*

Butterflies exhibiting Batesian mimicry

Batesian mimicry

Unlike Mullerian mimicry, not all the mimics are distasteful, most species being chemically unprotected and gain protection from their resemblance to distasteful ones. From this it follows that the mimics must never exceed a certain number relative to the models' number or the protection will be lost. This phenomenon has been built up from those variants that confer high survival value where predators learn to shun well-marked unpalatable types and natural selection pressure is exerted in favour of the closest resemblance to those types. Bates observed this initially in Africa by studying the widespread African butterfly, *Papilio dardanus*. The males of this species do not mimic and are therefore conspicuous. The females mimic many unpalatable species of models and only in Madagascar and Abyssinia resemble the males.

Light form favoured on lichen-covered trunk

Natural selection

With modern knowledge in genetic, ecological, zoological, and mathematical fields, the effects of an environment on a certain variant of a species can be measured. This enables a researcher to study long-term effects and find out if evolutionary change does take place.

H. B. D. Kettlewell's research on 'industrial melanism' in moths illustrates evolutionary changes. This is the phenomenon in which moths are changing their complicated patterns from a light to an all-black coloration. The first work by Kettlewell was with the Peppered Moth (*Biston betularia*), but this is not the only species which is changing.

Before the Industrial Revolution the majority of trees throughout Great Britain had lichens on their trunks. Now such trees occur only in unpolluted areas. Carbon dust has killed the lichens on trees in industrial areas and rendered their trunks and branches black. The typical grey form of the

Melanic form favoured on sooty trunk

Peppered Moth is well camouflaged for resting on the lichen-covered bark of trees, being almost invisible to predating birds. Before 1848 this was the only form recorded, but soon after this a dark melanic variety *Biston carbonaria,* appeared. Due to a genetic mutation this variety differs from *betularia* by a single recurring dominant gene slightly more vigorous than the normal grey type. In the mid-Victorian days because it was more conspicuous, the *carbonaria* variety was constantly eliminated. However, the gradual darkening of the moths' environment meant that the more conspicuous light *betularia* variety was taken by birds rather than the dark *carbonaria*.

By 1900 melanic *carbonaria* outnumbered the *betularia* form in the Manchester area. There had been a sweeping change in just fifty years. Today, in only the unpolluted areas of Northern Scotland, West Devon, and Cornwall, and parts of the south coast is the *betularia* variety found.

Natural selection in the land snail

The land snail (*Cepaea nemoralis*) is highly polymorphic and the many colour variations and banding on the shell have been studied in the snails' various environments by Cain and Sheppard. The shell can be yellow, pink, or brown and can also be five-banded, one-banded, or unbanded. All combinations of shell colour and banding pattern can be found but some are much commoner than others. Brown is dominant to pink which is dominant to yellow, the three being apparently controlled by three allelomorphs. Unbanded is dominant to the presence of bands and the locus is linked with that controlling colour. The one-banded form is dominant to five banded and is not controlled by the locus which determines the presence or absence of bands. The proportions of the various colour and banding types differ greatly from one colony to another and Cain and Sheppard found that they are strictly related to the ecology of the site. The commonest shell colouring is that which is least conspicuous, against background vegetation, to predators. The most advantageous shell colours are yellow (greenish when the animal is inside) in green areas, pink on leaf litter, and reds and browns in beech woods with red litter and numerous exposures of blackish soil.

Migration of snails from unsuitable to suitable areas is highly unlikely as distances would be prohibitive and so it appears that the appearance of the snails in certain areas results from the action of predators hunting by sight. The predators include small mammals such as the rabbit, which may select by tone as most of the mammalian species are colour blind. Birds, especially the song thrush, are the chief enemies as they can distinguish colour, and they smash shells on convenient stones, 'thrush anvils', to obtain the soft body. By studying the shell remains it can be decided whether there is any colour selection by the thrush, or whether it eats snails at random. Much study in different areas has shown that the birds capture snails selectively, destroying an unduly large proportion of those whose colour and patterns match their habitat least well. Here again natural selection is in action, although in certain areas selection for colour pattern appears absent due to the wide range of specimens found.

Thrushes are the main enemies of the land snail.

A selection of shell patterns of *Cepaea nemoralis*

COURSE OF EVOLUTION

The long period from the origin of the Earth, about 5,000 million years ago, to the beginning of the Cambrian period, 520 million years ago, is often referred to as the barren past. However the pre-Cambrian period definitely supported certain unspecialized principal forms of life, as the Earth probably became habitable some 3,000 million years ago. Very few fossils can be found from the deeper pre-Cambrian rocks, although the Cambrian rocks reveal a wide and diversified account of animal and plant life.

Several plausible explanations are available, however, to account for this lack of fossil evidence. Many of the fossil forms may have been destroyed by erosion or by metamorphism during the passage of time, or they may be as yet undiscovered. However, an important point is that the early forms were probably soft bodied, and therefore incapable of preservation as fossils.

The earliest fossils known to us are still extraordinarily complex in comparison with the lowly organisms which must have existed at the dawn of life. Pre-Cambrian fossils

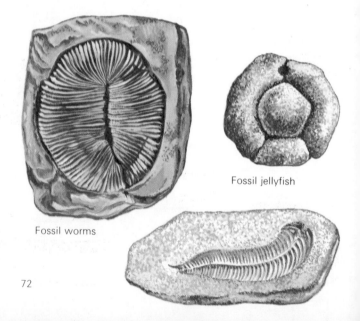

Fossil jellyfish

Fossil worms

A middle Cambrian scene

tend to cause most confusion to the palaeontologists as regards
their age and their actual form. A medusoid impression that
probably represents a jellyfish has been discovered in pre-
Cambrian rocks of the Grand Canyon, some algae or sponges
identified from Scotland's pre-Cambrian rocks, and important
evidence of worms, coelenterates and echinoderms from
Australia.

One often hears of the large gap between the pre-Cambrian
period and the lower Cambrian period when apparently there
was a sudden appearance of a wide range of well fossilized
plant and animal life. It must be remembered, however, that
the Cambrian period lasted some 90 million years, so there was
time enough for apparent 'sudden' forms to occur. There are
abundant records of marine invertebrates and algae but trilobites appear to have been the dominant group. In fact there
may have existed even in this time representatives from all the
phyla of plants and animals.

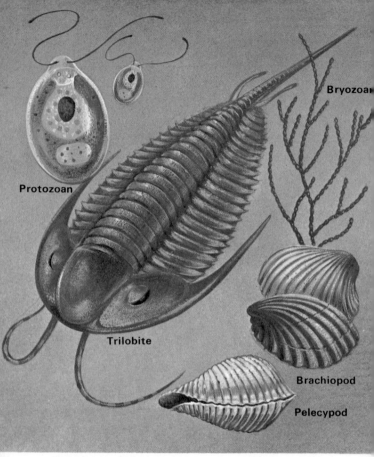

Bryozoa

Protozoan

Trilobite

Brachiopod

Pelecypod

A diverse number of invertebrate forms evolved in the shallow seas of the Paleozoic and Cambrian periods.

Marine invertebrates

One must remember that the Earth was far different at the beginning of the Palaeozoic era than it is today. An enormous land mass appears to have stretched then from the Galapagos Islands, west of South America, to western Australia. The climate was equable but the land was barren except for patches of green algae encrusting the rocks. In the seas, life had not evolved beyond the invertebrate fauna. However, even among these animals there was tremendous radiation into a diverse number of forms.

Present in the shallow seas were sponges, jellyfish, gastropods, brachiopods, worms, arthropods, and echinoderms. The sponges (phylum Porifera) are the simplest of all the multi-cellular animals. The Cambrian sponges were fixed in position and varied widely in size, shape and colour. Until 100 years ago their actual place in the classification table was undecided. They were finally shown to be animals, probably arising from protozoans. The body of a sponge is more or less a sieve made up of animal cells. The body wall is perforated throughout by pores and canals and water flows into the central cavity and out through the top by one or more openings. Many are supported by silica spicules, either separate or joined to form a skeletal framework. Often these have been fossilized, producing most beautiful patterns.

Above these fixed dwellers on the sea floor floated the jellyfish polyps. The soft bodies were rarely preserved, but even so they are known from early Cambrian times. Sea anemones, corals, sea fans, and sea pens all belong with the jellyfish in the phylum Coelenterata. All are aquatic, either solitary or colonial, and have a sac-like body cavity with a ring of tentacles surrounding the single opening. They feed on a wide range of life, some capturing and eating only microscopic organisms but other more formidable ones devouring molluscs, crustaceans, and fish which are first paralysed by means of stinging cells.

The most characteristic of all forms of life in the Cambrian seas were the brachiopods, the bottom-living marine 'shellfish' and the ruling trilobites. Today, there are only about 200 species alive in the seas but in the lower Palaeozoic they were very abundant and universal. The brachiopod shell consists of two valves, bilaterally symmetrical, surrounding and protecting the soft parts of the body. Instead of the two valves matching each other they differ in size, unlike a true bivalve shellfish such as the cockle or mussel. The animal is usually attached to the ocean floor by a long, fleshy stalk or pedicel, which usually perforates the larger valve. *Lingula*, a small tongue-shaped species, is an example of a simple brachiopod present in Cambrian seas that has remained unchanged.

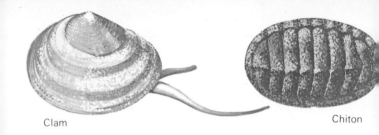

Clam

Chiton

Molluscs

Slow-moving snails found in the rock pools of Cambrian times were the oldest representatives of one of the most varied and successful groups in the world today. The phylum Mollusca includes animals such as slugs, snails, oysters, clams, chitons, cuttlefish, and octopuses. There is a great variety in size, form, and shape but all share a fundamentally similar body plan. They have a soft unsegmented body, an anterior head and a large visceral mass supported on a fleshy muscular foot. Surrounding most of the body is a thin fleshy layer, the mantle, which secretes a shell in most members of the phylum. Snails have a single coiled shell, while in the clams and oysters the shell consists of two valves. The chiton has eight protective movable plates, while in cuttlefish and squids the shell is completely internal.

Although the early forms were entirely marine, as conditions changed the gastropods slowly evolved into a most widely adapted and diverse group. Today the 20,000 modern species inhabit a very wide range of environments: marine, freshwater, terrestrial, arctic, tropic, desert, and from ocean depths to over 18,000 feet above sea level.

Members of the phylum Mollusca (*top* and *bottom*) and the evolution of various invertebrates (*opposite*)

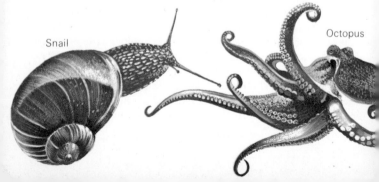

Snail

Octopus

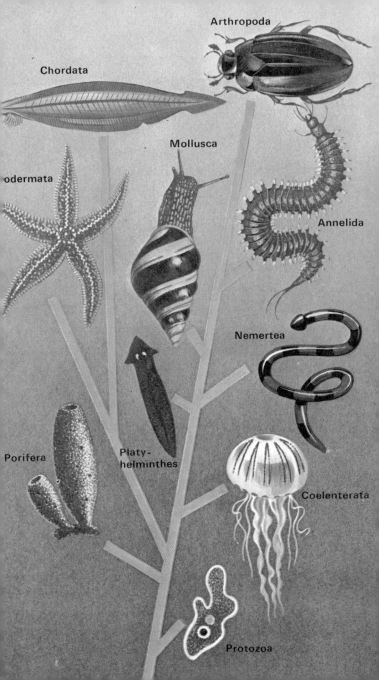

Chordata

Arthropoda

Mollusca

odermata

Annelida

Nemertea

Platy-
helminthes

Porifera

Coelenterata

Protozoa

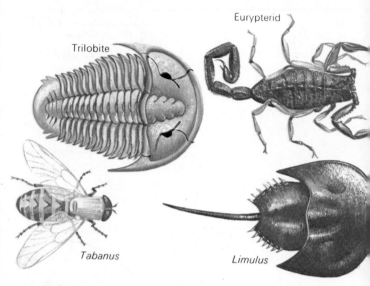

Primitive (*top*) and more modern representatives (*bottom*) of the phylum Arthropoda

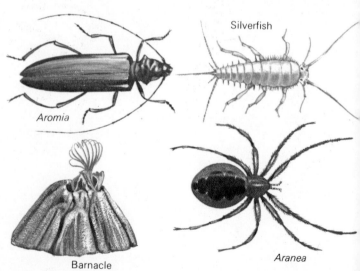

Arthropods

The Arthropoda contains about three-quarters of all known animal species including shrimps, crabs, lobsters, barnacles, spiders, scorpions, ticks, and insects as well as the extinct trilobites and eurypterids.

Dominating the Cambrian seas were the trilobites, and although some 1,000 genera were alive in that period, today they are completely extinct. The woodlice-like appearance is perhaps one of the most familiar forms of all fossils. Trilobites underwent adaptive radiation evolving many forms suited to particular environments. They started a steady decline in the Ordovician until they had dwindled to extinction in the late Permian times, some 225 million years ago.

The extinct eurypterids are the ancient aquatic relatives of the scorpions. Their usual length was about a foot though some reached 9 feet in length. Scorpion-shaped, the oldest eurypterids originate from the Ordovician, reaching their peak in the Silurian and the Devonian and becoming extinct in the Permian.

Also related to the eurypterids are the living horseshoe crabs, whose earliest ancestors appeared in the Cambrian. *Limulus*, the King Crab, is often spoken of as a living fossil, and although only marine forms occur today, they were the first animals to colonize fresh water.

The complete colonization of freshwater and land could take place only after colonization by plants. It is therefore no accident that terrestrial animals and plants appear at roughly the same period of geological time. Plants provided the food for the first terrestrial animals which were arthropods. The interdependence of plants and animals is of great evolutionary importance in the appearance of certain groups of animals and for the decline of others, as the flora of the various periods changed.

It is amazing how successful the orders of the arthropods have been in the conquest of the land, and for this certain characteristics are responsible. Extremely important is the chitinous shell which protects them from the hazards of desiccation. Furthermore, arthropods were already equipped with appendages which could support their bodies when out of the water, and provide an efficient means of locomotion.

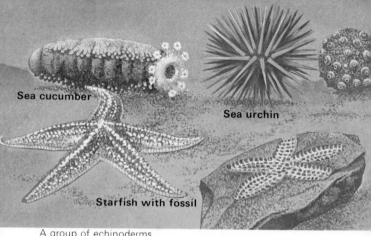

Sea cucumber

Sea urchin

Starfish with fossil

A group of echinoderms

Echinoderms, annelids and bryozoans

The phylum Echinodermata includes the living starfish, sea urchins, brittle stars, sea lilies, and sea cucumbers, as well as the extinct blastoids, cystoids and edrioasteroids. The earliest forms date back to the Cambrian period, and the phylum is entirely marine although some are free-living while others are sedentary. They all tend to be spiny, and the body is encased either by a calcareous roundish test as in the sea urchins, or by a leathery skin having a star shape, studded with small calcareous plates as in the starfish. Usually a five-fold symmetry is apparent. The larval forms show a striking resemblance to the larvae of hemichordates and this has led to linking of the two groups by some zoologists. The echinoderms may be close to the forms from which the chordates evolved.

The Cambrian echinoderms are all representatives of archaic and now extinct groups. *Edrioaster*, an extinct edrioasteroid, was globular, with five sinuous arms radiating from the mouth. Eocrinoids living attached to the sea floor in the Cambrian were probably ancestors to the later crinoids and cystoids. By the middle Palaeozoic the familar starfish, sea urchins, sea cucumbers, and crinoids, as well as a number of extinct forms had evolved. The most important of the Palaeozoic echinoderms were the crinoids and the extinct blastoids, both of which first appear in the Ordovician. Crinoids were common in limestone deposits from the Silurian onwards. Some were free swimming, others fixed by a stem.

Included in the phylum Annelida are the marine bristle worms, the Polychaetes, the most generalized class of annelids. The earthworms and relatives are grouped in the Oligochaeta (few bristles). The leeches, class Hirudinea, are also annelids. In Cambrian times only marine worms had evolved but they seem to have been as widespread as the marine worms of today. The soft bodies were not suitable for fossil preservation, but nevertheless evidence of their form and habit is often found. The worms are usually represented by fossil tracks, tubes, burrows and parts of their chitinous jaws (scolenondonts).

Bryozoans were not common until the Ordovician period although geologists do seem to have neglected this group. Many bryozoans resemble corals in the external form as the individual animal secretes a calcareous mat-like or frond-like skeleton. Bryozoa is a name meaning 'moss animals' and refers to the plant-like appearance of many bryozoans.

Representatives of the Annelida and Bryozoa

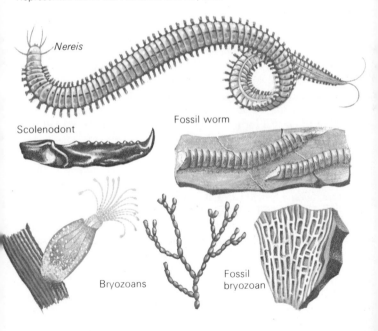

Nereis

Scolenodont

Fossil worm

Bryozoans

Fossil bryozoan

Origin of the vertebrates

During the Silurian there were a number of large land masses with an equable climate surrounded by warmish seas, where limestone, sand and mud could slowly accumulate. The invertebrates were still masters of the sea, but it was during these times that the vertebrates and land plants emerged.

A vertebrate has a hard, narrow supporting structure extending along the dorsal surface of its body. In most characteristic groups this is made of true bone and segmented to give rigidity and flexibility. We classify animals with this backbone into the phylum Chordata.

The evidence for the evolution of the vertebrates comes from the study of embryos and by comparative anatomy. Clues are found by investigating a few peculiar vertebrates which at first sight bear little resemblance to the higher backboned species. They are known as the Protochordata, which includes the sea-squirts or ascidians, lancelets and acorn worms. They are classed with the vertebrates because they posses, at some time in their life history, a stiffening rod (the notochord), pharyngeal gill pouches or slits, and a dorsal tubular nerve cord.

The tunicates or sea squirts in their adult form seem at first wrongly classified. They are sedentary, sac-shaped almost sponge-like, with a stiff outer body, often colonial, clinging to rocks and seaweeds in shallow seas. They do not have a nerve cord, vertebrae or a notchord, although they have gill-slits. However, the sea squirt is a vertebrate and the evidence comes from the larval form. The larva is free-swimming and tadpole shaped, with a notochord and a nerve cord.

The sea squirt larva closely resembles adult Amphioxus, the lancelet. Amphioxus is another early chordate, about 2 inches in length, translucent and streamlined in shape. It too has a notochord, a dorsal nerve cord, and gills with a simple ventral digestive tract.

Although these modern chordates have undoubtedly changed from the original unspecialized forms, they are not too different from the early chordates, which provide a link between the invertebrates and the vertebrates.

Protochordates provide clues to the origin of the vertebrate form.

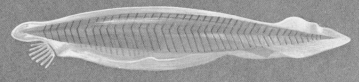
Amphioxus

Larva of sea squirt

Sea squirt colony

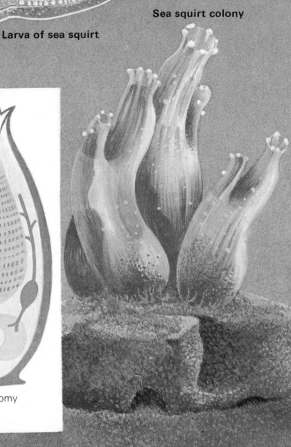

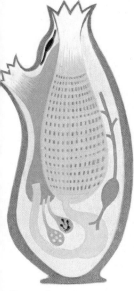
Sea squirt anatomy

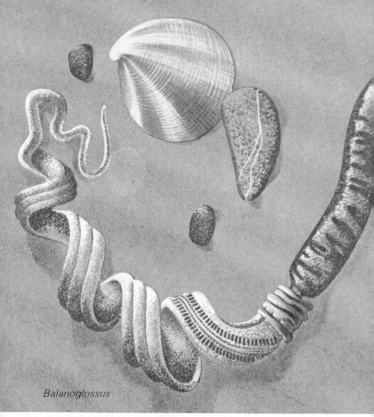

Balanoglossus

A vital link – the acorn worms

The acorn worms are worm-like, simply-organized burrowing marine worms including *Balanoglossus*, *Glossobalanus*, *Ptychodera* and *Saccoglossus*. They are the most primitive of vertebrates and in external appearance resemble invertebrates rather than vertebrates. Found in burrows in the sands and muds of shallow marine waters, they feed by extending the mouth from their burrow, and drawing in particles of sand and mud down the simple alimentary canal which extracts organic matter. The mouth is not to be mistaken for the proboscis, a muscular, cylindrical burrowing structure which retracts into a collar surrounding the mouth.

The identifying chordate structures are a dorsal nerve cord

and a shorter ventral one, gill slits and a notochord. Again it is the larvae that are of extreme importance. They are small, bell-shaped organisms which drift with the currents of the sea. Around the mouth is a ring of cilia which changes shape as the larva develops. The larval form is termed a *tornaris* larva. This primitive chordate larva when compared with the larvae of worms and molluscs shows similarity, but only the molluscs have a single ciliated ring encircling the body in front of the mouth. It is feasible that larvae such as these may have given rise to the later vertebrates. However, by which route and from which invertebrate group the vertebrates evolved needs further investigation. Certain authorities state that arthropods and annelids provide the link due to them having a gut and nerve cord. However, the gut in the arthropods is dorsal and the nerve cord ventral, so to produce a characteristic vertebrate organization, they must be reversed. A complete inversion of two complex parts seems unlikely. The current view is that the most likely ancestors of chordates are the echinoderms. Here again it is the larval echinoderm which links the invertebrates quite definitely with the lower chordates. The larvae of echinoderms closely resemble the larvae of the acorn worms.

Research has also shown physiological and anatomical similarities. The evidence points to a common ancestor for the echinoderms and chordates which probably became extinct and for which no fossil evidence has yet been found.

Larvae of an echinoderm (*left*) and *Balanoglossus* (*right*)

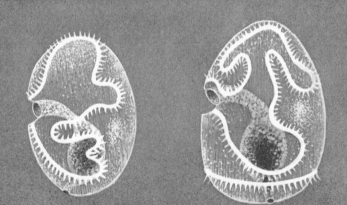

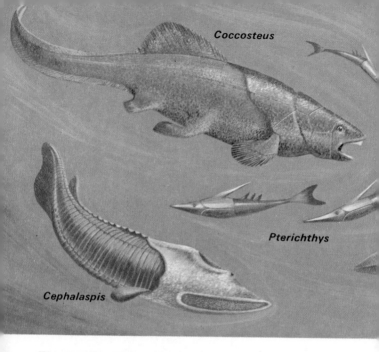

Coccosteus

Pterichthys

Cephalaspis

The earliest vertebrates

The evolution of some vertebrate larval forms to primitive fish-like forms can hardly be expected to have left traces in rocks, as they both most probably were soft bodied. The earliest traces are from the middle Ordovician period, and some body fragments have been discovered in the Colorado and Wyoming area. There are four classes of fishes identified today, all having gills, skin with scales, and fins: the Agnatha, the Placoderms, the Chondrichthyes and the Osteichthyes. All had evolved by Devonian times. The most primitive are the Agnatha, which are the only craniates lacking true jaws and paired fins. The Devonian period saw a great expansion of the fishes, and much fossil evidence has been discovered to give good coverage of groups that had evolved.

Typical of the Agnatha group and found in the Devonian is *Cephalaspis*. As with most cephalaspids it had a flattened, bony head shield and a scale-covered body. Cephalaspids appear to have marine and freshwater members, probably all were

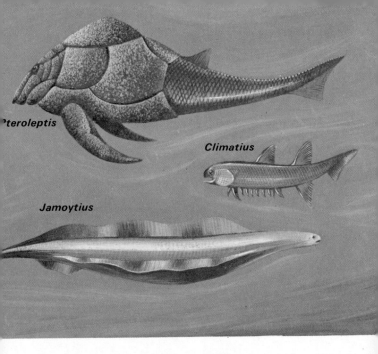

Pteroleptis

Climatius

Jamoytius

bottom dwellers, grubbing in the mud of streams and ocean bottoms. *Jamoytius* is torpedo-shaped and bears median lateral fin folds, with large eyes. Superficially it is not too unlike Amphioxus.

The placoderms became extinct in the upper Silurian but they had developed primitive jaws and paired fins, and radiated into a variety of forms. The development of jaws and fins are of extreme importance and represent a landmark in the evolution of the whole vertebrate kingdom and not only in the fish phylum. Until the appearance of jaws, animals were confined to the mud-grubbing life, but with jaws new food sources became available together with new aquatic environments. Without bony fins animals could never have left the water for a terrestrial life.

The Placoderms probably originated from primitive Agnatha and although the whole group marked a major step in the evolution of the vertebrates, they were reduced by the end of the Devonian, and extinct by the close of the Palaeozoic.

Bony fishes—Osteichthyes

Bony fishes are the most abundant, diverse, and complex group of fishes. They appeared in mid-Devonian times and flourished greatly in the Palaeozoic, by this time almost in sole possession in fresh water lakes and streams. They have a bony skeleton, and scale-covered bodies. Some fossil fishes and a few living fishes have lungs; the rest have a swim-bladder which controls buoyancy.

From the outset there appears to have been two distinct types of bony fishes. The first of these is the actinopterygians, or ray-finned fishes, the four orders of which are the palaeoniscids, the chondrosteans, the holosteans and the teleosts. Most have died out, but the teleosts are dominant today.

The second smaller division of the bony fishes in the Choanichthes, containing air-breathing fishes with internal nostrils opening into the mouth. There are few living representatives but their importance lies in tracing the evolutionary course of terrestrial vertebrates. The two orders are the Dipnoi and the Crossoptygii.

Primitive fishes of the Devonian

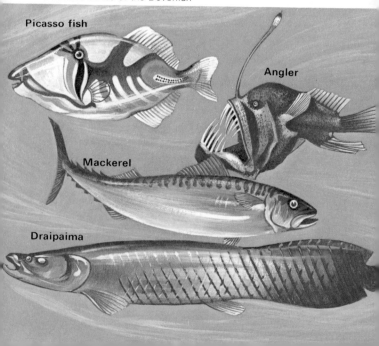

Picasso fish

Angler

Mackerel

Draipaima

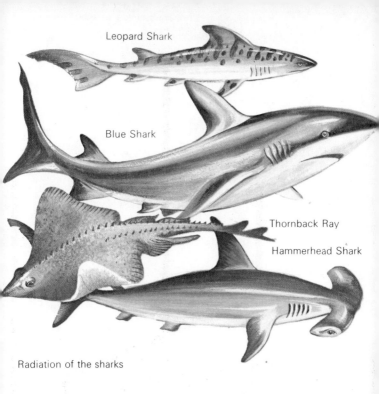

Leopard Shark

Blue Shark

Thornback Ray

Hammerhead Shark

Radiation of the sharks

Cartilaginous fishes — Chondrichthyes

Sharks, rays, and chimeras are classed in the Chondrichthyes. They are cartilaginous, predatory fishes and usually powerful swimmers. They depend on their large nostrils to track down their prey, rather than on their eyes which are quite small. The sharks appeared in the Devonian and underwent considerable radiation, becoming adapted to all types of habit by the upper Palaeozoic. Many became extinct by the end of the Permian, but even so there are about 600 successful species distributed widely through the oceans of the world. Fossil evidence shows that the early representatives were remarkably like the modern living species. *Cladoselache*, a well known fossil from the upper Devonian of Ohio, is some 3 feet in length with well developed, broad fins and a streamlined naked body.

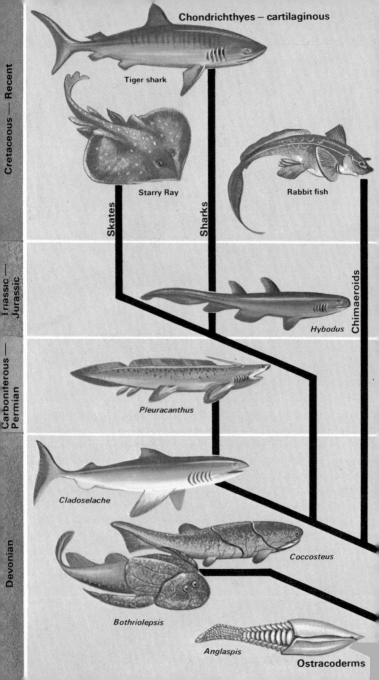

Chondrichthyes — cartilaginous

Tiger shark

Starry Ray

Rabbit fish

Skates

Sharks

Chimaeroids

Hybodus

Pleuracanthus

Cladoselache

Coccosteus

Bothriolepis

Anglaspis

Ostracoderms

Cretaceous — Recent

Triassic — Jurassic

Carboniferous — Permian

Devonian

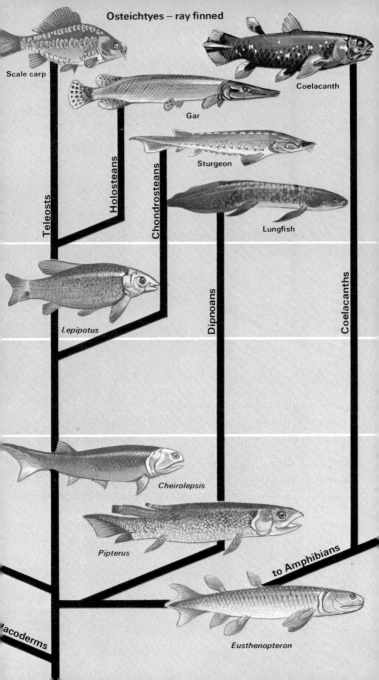

Osteichtyes – ray finned

Scale carp

Coelacanth

Gar

Sturgeon

Lungfish

Teleosts

Holosteans

Chondrosteans

Dipnoans

Coelacanths

Lepipotus

Cheirolepsis

Pipterus

to Amphibians

Placoderms

Eusthenopteron

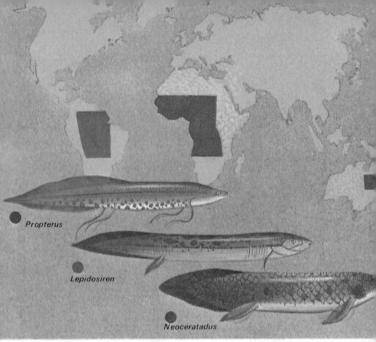

Propterus

Lepidosiren

Neoceratadus

Distribution of the lungfish

The lungfish or Dipnoi

Three genera only of lungfishes survive in the tropics:-
Neoceratadus in the Burnet rivers of Queensland, Australia;
Protopterus in the White Nile, some of the great lakes and the
Congo region of Africa; and *Lepidosiren* in the Amazon and
Parana rivers of South America.

The lungfishes possess some of the features that enabled
some of the evolving relatives to emerge from the water and
become terrestrial vertebrates. The critical evolutionary period
took place in stagnant swamps and drying pools, where the
waters were low in oxygen and subject to periodic evaporation. The lungfishes, possessing lung structures, became pro-
ficient in gulping down oxygen directly from the air at the
surface. The early forms of lungfishes may well have branched
to give rise to amphibians, and indeed the larvae of *Lepidosiren*
and *Protopterus* shows similarity to those of amphibians, by
having suckers and external gills.

The Coelacanth

The Crossopterygii are a group which became rare in the Carboniferous and apparently disappeared after the Permian. Knowledge of these lobe-finned fishes was quite limited as no members had apparently survived. From fossils it was evident that they had lungs and internal nostrils and in nearly every respect resembled the primitive amphibian skeleton. The fishes lacked legs of course, but, their fins contained a fleshy lobe, within which were bony skeletal supports. The basic pattern of these fishes is therefore comparable to that of a land mammal. The early typical lobe-fins lived in the fresh-water areas and became extinct quite early on. No later fossils were known, so it was assumed that the crossopterygians were entirely extinct.

However, the amazing discovery in 1939 of a coelacanth fish off East London, South Africa, gave evidence that for more than 70 million years coelacanths had survived, comparatively unchanged. Several specimens have now been caught, so extensive knowledge of the fish is available.

The 'living fossil' Coelacanth

Polypterus

Actinopterygii—ray-finned fishes

The Actinopterygii fins are composed of a web of skin supported by horny rays only. Flesh and bone is confined to the base of the fin. Many marked differences distinguish this group from the Choanichthes, including scale structure, different patterns of the bones of the head and lack of internal nostrils. The eyes tend to be large and seem to be the dominant sense organ, the sense of smell is of relative unimportance. The lung is transformed into a hydrostatic organ, the swim-bladder. The number of eggs laid is far higher than any other fish group, and although small in size they number up to thousands and even millions from one female alone. Quantity and not quality is the principle and although many may die in early life, numbers are quickly renewed, probably an important factor in the success of the group.

The palaeoniscids are the oldest ray-finned fishes, appearing in mid-Devonian times. At that time they were greatly outnumbered by the lungfishes and lobe-finned fishes. Two African fishes *Polypterus* and *Calamoichthyes,* represent the earliest ray-finned stage, although in a modified form.

Bowfin

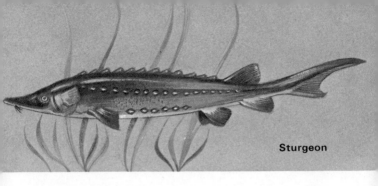

Sturgeon

Sturgeon and paddle fishes—chondrosteans

Found in the Mississippi river are another two survivors from an early stage of the Actinopterygii evolution. The sturgeon and the paddlefish are rather more advanced, with a typical air bladder rather than a lung. However, the fins are more primitive than in *Polypterus*. The tail fin is shark-like, and the scales have degenerated to a stage where they are found only in a series along the length of the body with tough leathery skin in between. With feeble jaws, small mouths and long snouts they are mud grubbers.

Bowfins and gar-pikes—holosteans

The holosteans emerged in Permian times from the palae-oniscid stock which they largely replaced. They had quite deep bodies, and a more advanced symmetrical homocercal tail. The only survivors are found in the Great Lakes of North America, and are completely successful in spite of the competition from higher bony fishes. The bowfin (*Amia*) or freshwater dogfish, and gar-pike (*Lepisosteus*) differ quite markedly in external features.

Garpike

Dragon fish

Sea Horse

The modern dominant teleosts

Arising during the late Triassic from holostean stock were the teleosts, comparatively rare until the Cretaceous by which time several lines of evolution had already begun. Today, some 20,000 bony fish species abound in the seas and waters to illustrate the success of the group.

The teleost plan of evolution has allowed for the development of a great range of specializations, fitting the animals to all sorts of situations in sea and freshwater. It is true that sea and freshwater have been in existence relatively unchanged throughout the course of the fishes evolution. One might ask why there was a need for the fishes to evolve new forms to suit apparently 'new' habitats. In a sense the fishes did not evolve to fit a new environment, but rather they found endless new

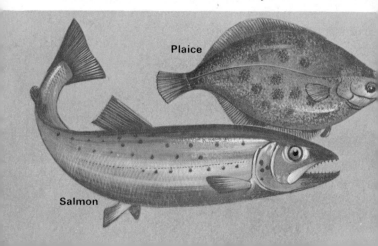

Plaice

Salmon

ways of living in the same aquatic environment.

The teleosts vary exceedingly in every respect. The more primitive body shape and general organization is shown by the Herring and Sardine, with slightly more advanced forms shown by the Trout and Salmon. Fin rays are more numerous and flexible in the primitive types, whereas later the fins are supported by only a few stout, movable spines. The body form takes on such tremendous variation that it seems in some comparisons that the fish concerned could not possibly be related. The flatfish, such as the Plaice, Sole and Flounder, have evolved a form perfectly adapted for life on the sea floor. They all have very flattened bodies and during development the young fish turns itself through 90° and the lower eye migrates to the new upper surface.

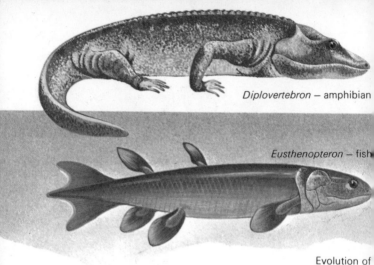

Diplovertebron – amphibian

Eusthenopteron – fish

Evolution of
amphibians
(*opposite*)

The conquest of land

The change from aquatic to terrestrial life by some of the
vertebrates was first initiated in the Devonian by the early
amphibians and completed by the reptiles in the late Palaeozoic.
Once more it was not a rapid process, but a slow progression
through various evolutionary stages. Certain bony fishes had
already developed essentials for land life, and for example,
some had developed primitive lungs. Furthermore, an
active life on land for any vertebrate demands that there is
good support for the body. In the lobe-finned fishes (e.g.
Eusthenopteron) the stout and fleshy lobe in the paired fins
gave possibilities for development into land limbs. All the
requirements necessary for terrestrial life were therefore
already initiated in the bony fishes.

Comparing the skulls of the earliest amphibians and those
of the Devonian Crossopterygii fishes, there can be little
doubt of the relationship between the amphibians and the fish
ancestors. However, transitionary fossils have still not been
discovered. Something of the transition is revealed by the
oldest fossil amphibia, the ichthyostegids from the upper
Devonian of East Greenland, showing features intermediate
between crossopterygians and the typical early Amphibia.

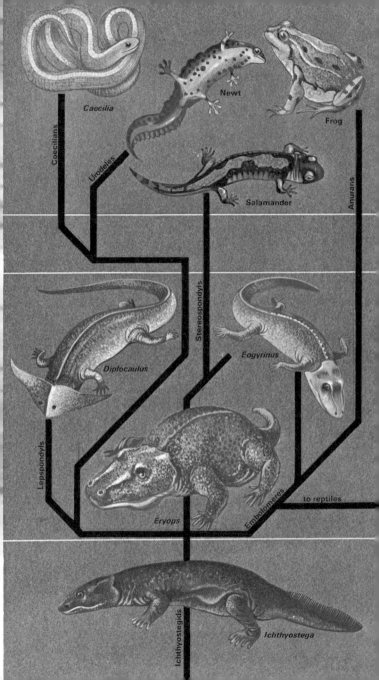

Caecilia

Newt

Frog

Coecilians

Urodeles

Salamander

Anurans

Stereospondyls

Diplocaulus

Eogyrinus

Lepspondyls

Eryops

Embolomeres

to reptiles

Ichthyostega

Ichthyostegids

Eryops

Amphibians past and present

Early amphibian fossils are rare, but excellent skulls and other skeletal parts have been found in Greenland, from the late Devonian. These amphibians of the Devonian and Mississippian periods, with their large heads, solidly roofed with thick bones, are stegocephalians, and show little resemblance to frogs and salamanders. The most conspicuous animals of the Pennsylvanian were the amphibians. They had strong legs and were able to move on land as well as in water. The giants in the water were the embolomeres, their limbs were reduced but they had powerful tails. *Eryops*, a large labyrin-

Salamander-like fossil larvae branchiosaurs

thodont was found by lakes and streams during the late Pennsylvanian and Permian. Although reptiles were becoming increasingly dominant these formidable amphibians were quite able to compete with them. Some developed armour so they could leave the water without fear of dehydration.

The amphibian form we know today, frogs, newts and salamanders, did not evolve until the Jurassic. Salamanders are similar in general appearance to the ancient types, but in the skull and skeleton there are many degenerate features. Their scaleless moist skin acts as an accessory breathing organ and to avoid desiccation they must remain in moist areas.

The tailless frogs are a flourishing group. Although they are often taken as examples of primitive amphibians they are

Salamander   Frog

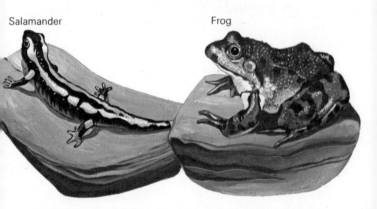

among the most specialized of vertebrates and this is attributable mainly to their hopping method of locomotion. They are well adapted for movement both on land and in water and can breathe in both environments providing they keep to moist areas when on land. They are also dependent on water for breeding and spend the cold winter months in hibernation. The frogs therefore illustrate well both the advances of the amphibians and the limitations of the amphibian way of life.

The Apoda, or limbless amphibians, are inconspicuous burrowing animals like worms found in the tropics. Their fossil history is unknown but they are a degenerate group, not closely related to the other orders.

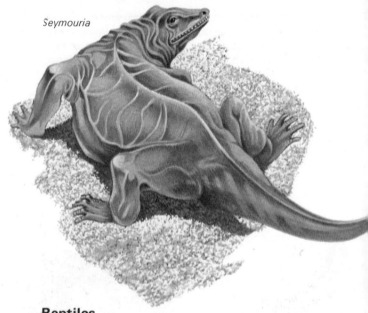

Seymouria

Reptiles

The amphibians, although the first to conquer the land, are a defeated group. Abundant at first, they are now an insignificant group among the tetrapod vertebrates. The reptiles have truly won the land and the reason is quite straightforward, when the development of the two groups is studied. Amphibians throughout their life are dependent on water; the eggs are always laid in water, and the adult must therefore periodically return (with the exception of the specialised caecilians, the Apoda).

With reptiles, no aquatic stage is necessary as they have evolved a type of egg which can be laid on land. The amphibian egg, developing in water, derives its oxygen, food and protection from this watery surround. The development of the reptile's egg (or amniote) provided the same benefits on land or within the mother's body, and the development of copulatory organs allowed internal fertilization.

From the early evolved reptiles, there branched numerous kinds, two paths evolving to give rise eventually to the warm-blooded birds and mammals. The early reptiles are known as cotylosaurs and referred to as the stem reptiles. Although the distinctive characteristic is the amniote egg, there are many other ways in which the reptiles differ from the amphibia. Many parts of the skeleton show differing structures and also circulatory and respiratory systems. The cotylosaurs probably developed from the seymouriamorphs. *Seymouria*, a squat creature about 2 feet in length, from the lower Permian of Texas, has given its name to the group. However, it is rather a misleading creature in its characteristics, some amphibian, others reptilian. The structure of its skull, teeth, and some features of its vertebrae are typically amphibian. Other characteristics of the vertebrae are reptilian as is the jaw suspension, the structure of the shoulder girdle and limbs. *Seymouria* itself is too recent to be the direct ancestor of the reptiles, but it is probably not greatly different from its ancestors. The question of which group it belonged to would soon be solved if it was known if it laid eggs on land or in the water. This does emphasize that it is the change in development which is the important feature in the history of reptiles.

Diometrodon

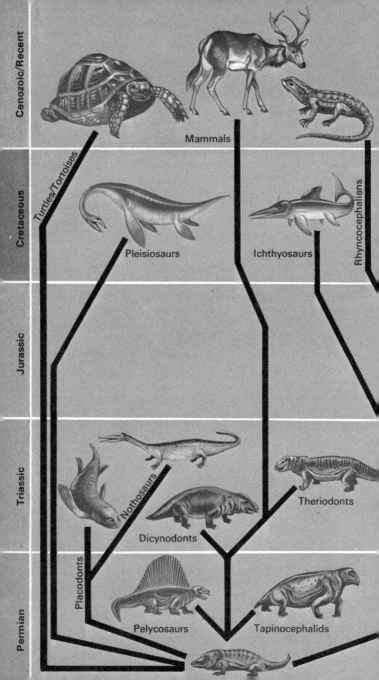

Cenozoic/Recent

Cretaceous

Jurassic

Triassic

Permian

Turtles/Tortoises

Mammals

Pleisiosaurs

Ichthyosaurs

Rhyncocephalians

Nothosaurs

Theriodonts

Dicynodonts

Placodonts

Pelycosaurs

Tapinocephalids

Lizards

Snakes

Crocodiles

Birds

Mososaurs

Ceratopsians

erosaurs

Ornithopods

Ankylosaurs

Theropods

Stegosaurs

Sauropods

Thecodonts

Evolution of reptiles

A reptilian Jurassic scene

The age of the reptiles

The age of the reptiles refers to the Mesozoic era, which was made up of the Triassic, Jurassic and Cretaceous periods, consisting of some 165 million years. Birds, mammals, and modern insects appeared for the first time, but the reptiles were the dominant group.

Early Mesozoic reptiles included a great variety of forms, some giving rise to the early ancestors of the ichthyosaurs and plesiosaurs which moved back to the sea. Among the terrestrial reptiles certain became bipedal, light-boned and fast-running saurians and gave rise to the dinosaurs. The cotylosaurs were the stem reptiles from which these later advanced groups radiated. The pelycosaurs evolved in the late Carboniferous, and were destined to play a most important part in the evolution of higher bertebrates. The pelycosaurs were a varied group, but most were rather sprawling creatures with long tails. Some exceeded 10 feet in length, and the most striking parts of these reptiles were their backs with greatly elongated neural spines, which appear to have been covered by skin and to have formed a sail-like structure. *Dimetrodon* was around 11 feet, a quite active predator with a large, deep skull and differentiated teeth. Its sail fin was about 3 feet high, but the function of this curious structure still remains a mystery. It may have been concerned with temperature control or even a form of protection.

It was from the pelycosaurs that the other great synapsid group developed, the therapsids, which became widespread in the Permian and Triassic. Some members are strikingly similar to the mammals, which descended from them.

The dicynodonts (dog-toothed) were a common and widespread Permo-Triassic group of large vegetarians, with reduced teeth. In the theriodonts (beast-toothed), the mammalian resemblances are most striking. *Cynognathus*, an active carnivore, is a typical member of this group.

Tiny and even more mammalian in character were the ictidosaurs or weasel lizards of the upper Triassic. The jaw articulation in some is of the mammalian type.

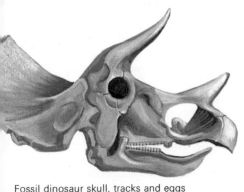

Fossil dinosaur skull, tracks and eggs

The dinosaurs

Although the Cretaceous saw the rise of the mammals it was the time of the mighty dinosaurs. The popular belief of dinosaurs is that of a single group of gigantic reptiles. Many reached 50 tons, but some were the size of hens. They were not a single group, but two distinct orders which had divided from the primitive stock. Early members of both orders were bipedal but several forms became tetrapods once more.

Members of the order Saurischia are identified by the structure of the pelvic girdle. The early members, such as *Conpscognathus* and *Ornitholestes*, were small and active, bipedal runners. The positioning of the toes was similar to those of perching birds, three forwards and one backwards. This led to the wrong identification of dinosaur footprints discovered in Triassic rocks, at first they were said to be prints of ancestral birds. This line produced by the end of the Cretaceous the largest carnivores, such as *Tyrannosaurus*,

Cretaceous dinosaurs

nearly 50 feet long and 20 feet high. They probably preyed upon the herbivorous dinosaurs and both became extinct, either because of climatic changes or competition with the mammals and birds.

An offshoot from this line gave some ostrich-like forms, *Struthiomimus* and *Ornithomimus*, walking on three toes with a grasping three-fingered hand. They were toothless, but developed a horny beak and probably ate eggs. Another line gave rise to the large four-legged herbivores: *Brontosaurus*, *Diplodocus* and *Brachiosaurus* were the largest of all terrestrial animals. *Brachiosaurus* from North American and East African deposits, was the real giant, with quite a small tail, some 80 feet long and weighing close to 50 tons.

The second group, order Ornithischia, are the duck-billed dinosaurs, often with 2,000 teeth and a beak. Most were heavily armed, such as *Stegosaurus* with spines on the back. *Triceratops* had huge horns and bony frill encircling the head.

Marine reptiles

The reptiles were the first group to firmly establish themselves on land, but they evolved some members which returned to the sea. Often when a particular group is well established in one habitat, it will also spread into other environments.

The reptiles that returned to the seas re-adapted to aquatic conditions and became as diverse as the shore reptiles. Six groups have been identified.

The turtles appeared in the Triassic and are little changed. They seem to have evolved from a reptile similar to *Eunotosaurus*. Land tortoises did not evolve until the Tertiary with the fresh-water terrapins and marine turtles radiating from them. In the Cretaceous some marine turtles grew to a length of 12 feet.

Highly adapted to an aquatic life were the ichthyosaurs or fish-reptiles, so called because of their very fish-like appearance. Their limbs were highly modified and they could

Jurassic marine fauna

therefore not come on to land to breed. It is suggested that the ichthyosaurs retained their eggs inside the body until they hatched; as do some snakes and lizards. Young ichthyosaurs have been seen within the body of an adult specimen and they must have swam in a fish-like manner using the tail as the main propulsive force. The oldest fossils are from the mid-Triassic and they continue until the late Cretaceous.

The most savage reptiles were the mosasaurs and geosaurs. The geosaurs were marine crocodiles of the Jurassic and lower Cretaceous. The mosasaurs seem to be confined to the Cretaceous, some were 30 feet long. The nothosaurs, the placodonts, and the plesiosaurs are all closely related, and paddled with their oar-like limbs. The early plesiosaurs, such as *Lariosaurus* from the Triassic, were about 3 feet long, but in the Jurassic and Cretaceous they were numerous and some were 50 feet long. Their jaws were highly modified with sharp teeth for crushing shellfish.

111

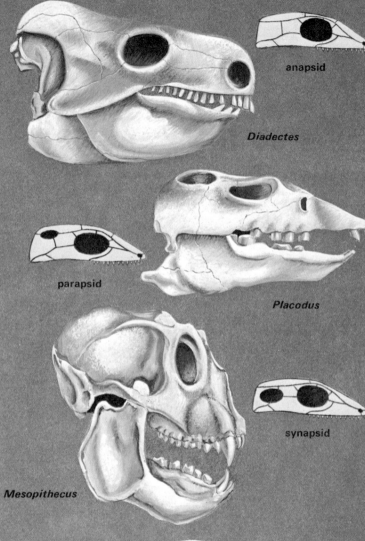

anapsid

Diadectes

parapsid

Placodus

synapsid

Mesopithecus

Goling

diapsid

Changes in reptilian skull structure

One very important feature of the reptiles is the evolution of the temporal region of their skulls. The structure of this region has given rise to the identification of four main groups and by close investigation linking relationships have been made between the groups. Even so although the temporal openings illustrate certain evolutionary trends, some zoologists think such a classification is misleading.

The most primitive reptiles, the cotylosaurs, had a complete bony roofing in the temporal region – the *anapsid* skull. The turtles (*Chelonia*) have a similar arrangement, although the bones at the back of the skull are reduced.

From this anapsid condition probably evolved two principal types. The *synapsid* group had a single opening (fossa) in each side bounded by the post-orbital and squamosal bones above, and the squamosal and jugal bones below. The reptiles with this type of skull were on the evolutionary line which led to the mammals. The skull was modified by a reduction in the bones and their positioning.

The other modification of the primitive anapsid condition gave the *diapsid* skull. Here, there are two openings in the temporal region. This type is found in many reptiles, including the crocodiles, the dinosaurs and the pterosaurs. A modification is found in the lizards and snakes. It is from the diapsid skull that the bird skull was modified. The birds lost the post-orbital bone so that the two openings in the temporal region and the eye form one large opening.

In other groups only a single opening and arch is present, similar to the synapsid condition. However, in these cases the opening is situated high on the skull, surrounded by the parietal bone above and squamosal and post-orbital below. This type of skull is termed *parapsid,* sometimes *euryapsid,* and is seen in the ichthyosaurs and plesiosaurs.

Although the classification of reptiles by their skulls is in some ways artificial, it does serve to indicate the main lines of their evolution and how the birds and mammals are descended from certain types.

Fossil reptile skulls reveal lines of evolution

Archaeopteryx

The origin of flight

The air was the last major environment to be colonized and in this the reptiles played an important part. For some 100 million years the insects had held the monopoly but from the Jurassic there are traces of flying reptiles. The pterosaurs or winged lizards showed the three essentials necessary for life in air – their skeleton was strong but light, they had wings and their sight was good due to the large cerebral hemispheres which control sight and muscle co-ordination. The Jurassic species were small with toothed jaws and long tails, but by Cretaceous times they often had a 25 foot wing span and long pointed toothless beaks, often longer than the vertebral column. They seem to have been gliders and soarers rather than active fliers, as the wings were less robust than in birds and bats.

The pterodactyls are more common in the Jurassic, than in the Cretaceous. They seem to have paralleled the evolution of the birds from a thecodont stock, but there is no evidence that they were feathered, although the wing had a membrane.

The second group was the evolution of the warm-blooded birds from the diapsid archosaurs. There is no detailed evidence of the transformation of cold-blooded reptiles into warm-blooded birds. However, certain fossils do give ideas of the intermediate stages. The oldest known birds are *Archaeopteryx* and *Archaeornis* found in the upper Jurassic of Bavaria, which show traces of feathers. Apart from this there is little to distinguish them from some of the dinosaurs.

During this time, birds underwent great changes and the fossils of the Cretaceous showed them to be true birds. A flightless swimmer *Hesperornis,* was a wingless, streamlined bird with a long beak and webbed feet. It probably plunged beneath the surface in search of prey. Another species *Ichthyornis,* was a powerful flier probably feeding on the shoreline or swooping down to pick up fish at the surface.

The origin of flight is still open to much discussion. One line of thought is that the ancestral birds were tree dwellers, and the developing wings broke falls. These leaps subsequently became glides. Another theory is that the ancestors were terrestrial, and feathered wings and tail developed to increase speed over the ground until they eventually lifted into full flapping flight.

The enormous diving *Hesperornis*

Trogons
Manakins etc
Thrushe
Nightjars etc.
Swifts
Ant Thrushes
Starlings
Bee-eaters
Toucans
Kingfishers
Puffbirds
Rollers
Hornbills
Woodpeckers
Lyrebirds
Grouse
Typical Owls
Hawks etc.
Barn Owls
Grebes
Ospreys
Cuckoo
Emus etc.
Hesperornis
Ostrich
Archaeopteryx

The radiation of bird evolution

The basic bird structure has become modified to produce the great variety of modern birds. Due to the few fossil remains, the direction of change has been built up from the study of present day forms. The radiation took place in the Cenozoic. The widely differing habitats of the birds range from the powerful birds of prey, the penguins, the waders of the shore

and estuary, to the game birds. The smallest bird is the Cuban Bee Hummingbird weighing 0·07 ounces; the ostrich is the largest and may weigh 300 pounds. The largest of all now extinct birds were the ten-foot high elephant birds of Madagascar. The vast number of genera are grouped into over forty orders, and the Passerines or perching birds comprise about half of all the species.

Phororhachos

Diatrymas

Moa

Dodo

Extinct flightless birds

The largest birds were all flightless. At first it was thought that they were all related but now this idea is not held. As they evolved terrestrial habits, their increase in size and rather ferocious nature protected them from their enemies.

The South American *Phororhachos* resembled *Diatrymas* closely. Some of their heads were horse sized, and fossils occur in the Oligocene, Meiocene, and Pleiocene rocks in Argentina.

Diatrymas was a formidable carnivorous ground creature of the early Eocene, some 7 feet high with reduced wings and a massive head.

The first bone of a Moa was brought to London in 1838 from New Zealand. Evidence shows that natives killed Moas for food and cooked them in earthen pits 5 feet in diameter. The largest birds were 12 feet tall.

The ugly Dodos were abundant for some time on Mauritius. The numbers of these swan-sized birds, related to pigeons, were reduced by Portuguese explorers in 1505 and later by the Dutch in 1598. By 1691 the Dodos were extinct.

Living flightless birds

Ostriches, rheas, emus, cassowaries and kiwis are collectively known as the ratites.

Ostriches live in the wild state in Africa and Arabia although fossil evidence shows ancestors were present in the Eocene in Switzerland. Their well-padded, two-toed feet enable them to run quite rapidly over stony and sandy deserts.

The rheas of the Argentine are sometimes called ostriches but they are not closely related. They have been found in South American Pleiocene formations.

The emu and cassowary group is restricted to Australia and New Guinea. The wings are even more reduced than those of the ostrich and rhea.

Relationships of the southern hemisphere penguins to other birds are not clearly understood. It seems they may have had a common ancestry with the oceanic albatrosses, shearwaters and fulmars, possibly in the late Cretaceous or early Cenozoic. Even the earliest forms seemed incapable of flight. Opinions differ as to how they evolved to become so beautifully adapted for an aquatic life.

Ostrich

Rhea

Emu

Penguin

The origins of mammals

The end of the Cretaceous marks the turning point for the reptiles, as the mammals began to replace them as the dominant vertebrates. There seems little doubt that the chief factors were the climatic and geographical changes which began in the early Tertiary. Temperate conditions replaced the previously widespread warm conditions. The cold-blooded reptiles lacked good insulation, they had become adapted to the warmth and were unable to adjust to the new conditions.

Climatic changes also affected the flora, the cycads had been the typical plants of the Mesozoic but these only survived in the warmer parts. They had provided the bulk of the herbivorous reptiles' food and as food became scarce a decline in the

Cumbersome herbivorous reptiles like *Ophiacodon* became extinct during the Cretaceous.

reptiles followed. The small size of the cranium may have also been a factor, or muscular movement and co-ordination may have become difficult with increasing size.

Mammals had been slowly evolving during the dinosaur dominance. Reptiles with mammalian features existed from early Triassic times, a typical example is *Cynognathus,* some 7 feet long. Outstanding mammalian features were differentiated teeth and limbs holding the body in a mammalian posture. It still had a long tail with reptilian articulation however.

The surviving mammals which lay eggs are the Australian

Echidna

Duckbilled Platypus (*Ornithorhynchus anatinus*) and the Spiny Anteater or Echidna (*Tachyglossus aculeatus*). They are so different from other mammals that they must have left the main mammalian stock in the early Mesozoic. Their organization may show us many of the characteristics of the Mesozoic mammals. Today they are highly specialized creatures, both are toothless as adults, although the young platypus has a few tooth rudiments. The absence of teeth is compensated for by the development of a broad, horny bill. The Echidna is protected by stout spines. It has powerful digging limbs and a long snout. Its mammalian characteristics include fur and it nurses its young but of course it lays eggs reptilian style.

Platypus

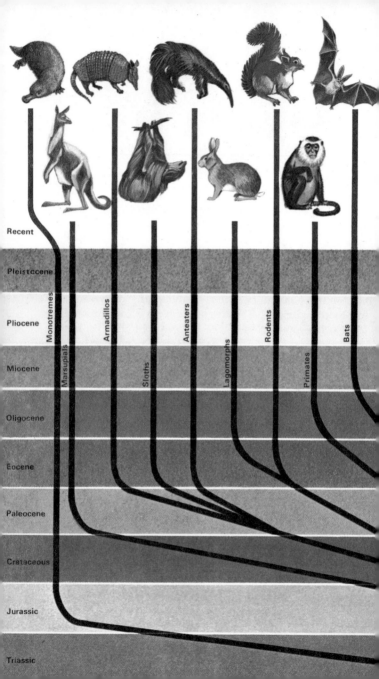

Recent

Pleistocene

Pliocene

Miocene

Oligocene

Eocene

Paleocene

Cretaceous

Jurassic

Triassic

Monotremes

Marsupials

Armadillos

Sloths

Anteaters

Lagomorphs

Rodents

Primates

Bats

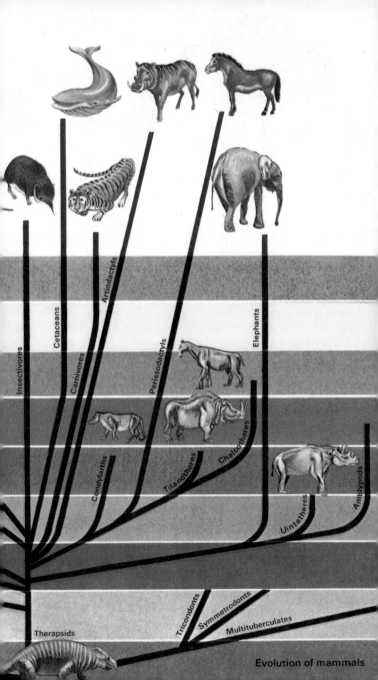

Insectivores

Cetaceans

Artiodactyls

Carnivores

Elephants

Perissodactyls

Condylarths

Titanotheres

Chalcotheres

Uintatheres

Amblypods

Therapsids

Triconodonts

Symmetrodonts

Multituberculates

Evolution of mammals

Bandicoot   Opossum

Pouched mammals – the marsupials

The pouched mammals show essential similar characteristics
to placental mammals but are primitive in many other
characteristics and undoubtedly diverged from the mammalian stocks at some early age. They portray quite well what
mammals were like in the late Cretaceous.

There are some 230 living species of marsupials, all varied
in habit and design, paralleling in the isolation of Australasia,
the adaptive radiation accomplished in other parts of the
world by the placentals. Today the majority are found in
Australia, with a few representatives in North and South
America, but in the Eocene they definitely occurred in
Europe and presumably became restricted by the competition imposed by rapidly evolving placentals.

The South American continent is now connected to the rest
of America by the narrow Isthmus of Panama. However,
during the Tertiary it was unconnected and the link was not
re-established until quite a late stage in the Cenozoic era. The

Tasmanian Wolf

Kangaroo

Koala   Phalanger

only marsupials found today in South America are opossums; some seventy-two species are recognized today. Fossil evidence shows that during the isolation, the marsupials radiated in several directions. Some paralleled the carnivorous cats and dogs of other continents, until North American connections became re-established, placentals invaded, quickly causing the extinction of the carnivorous marsupials and leaving only a few opossums.

Australian marsupials have survived because the large continent was isolated from the rest of the world in the Cretaceous and remains isolated today. Pouched mammals had entered before the isolation and it was only on man-made introductions that other land mammals entered (excluding bats and rats which apparently originated from the East Indies). Thus a curious and fascinating fauna was able to evolve and again many forms have paralleled groups of higher mammals evolved in other areas of the world.

Marsupials successfully radiated into a multitude of forms in habitats free from competition from placental mammals.

Tasmanian Devil   Wombat

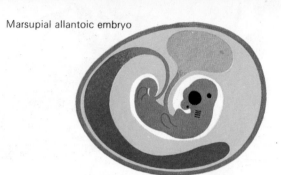

Marsupial allantoic embryo

Placental mammals

The evolution of primitive placentals marks the final stage in the general line of mammalian ascent, the placenta acting as an intermediate structure by which the developing embryo is nourished. The placental method of reproduction results in the birth of young at a much more advanced stage than in marsupials. The factor of longer prenatal development and the period of post-natal care and training which follows, together with the much larger and efficient brains which most groups possessed, are features of great significance in the dominance of the placental mammals.

The earliest placentals appeared in the Cretaceous, about the same time as the marsupials, which seems to point to a common ancestor. They were insectivores, the forerunners of the living moles, shrews and hedgehogs. Some show similar

The placenta enables the embryo mammal to develop to a more advanced state before birth.

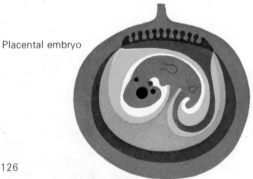

Placental embryo

Deltatheridium

structures and appearance, whereas others are definitely more primitive in form. All were small, mostly unobtrusive, nocturnal in habit and totally insignificant in comparison with the dinosaurian giants among which they lived, and the mammals which later evolved from these primitive placental stock beginnings.

Deltatheridium is a fossil insectivore found in the upper Cretaceous of Mongolia and shows characteristics very close to those of the ancestor not only of insectivores but of all placental mammals. Its skull was tubular and elongated, but lacked the specialized snout of some of the modern forms.

From these early mammals radiated various forms adapting and filling every suitable environment. Many of the mammals became larger and this may be connected with the development of a larger brain. The limbs became longer and specialized for locomotion. The teeth number reduced in many and their shape became suited to a particular diet.

Modern insectivorous shrews resemble their remote ancestors.

Flesh-eating mammals

The earliest Cretaceous mammals were most probably insectivorous and it is therefore not surprising that some of the descendants became flesh-eating. Indeed practically all modern flesh-eating mammals since have evolved from this single stock.

The major changes in mammals of carnivorous habits were in the design of the teeth. The carnivore kill is made using mainly the teeth and these have to pierce stout hide, cut tough tendons and grind hard bone. The fact that flesh is quite easily digested means that it need not be well chewed. Thus it is the front part of the mouth that is highly adapted. The front incisors are the biting and tearing teeth; the side canines are long and pointed, piercing and stabbing weapons in all the carnivores. The cheek teeth are usually reduced in number and those that are present have sharp ridges and pointed cusps rather than flat surfaces. In many flesh eaters, such as the dogs, a very specialized pair of teeth called 'carnasials' have been developed on either side of the jaw. The teeth do not meet directly but slide past each other, scissor-like, slicing or cracking very tough tissue.

The skeleton has a more primitive basis, for it must remain

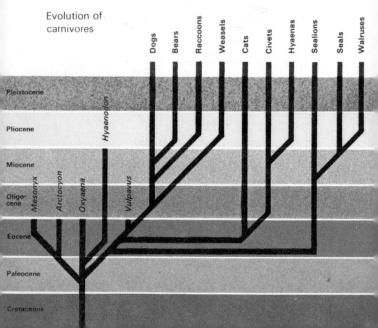

Evolution of carnivores

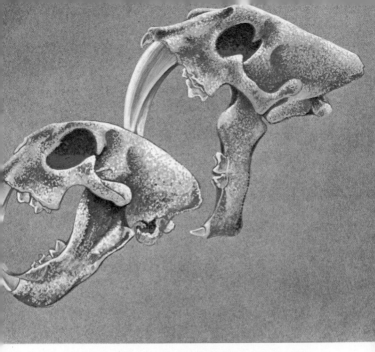

Skull of Sabre-toothed Tiger (*above*) compared to that of modern cat (*left*) to show modifications for striking and biting

supple to enable the animal to move fast. It does not require a reduced toe number or hooves as the claws are important in climbing trees, and for attacking and holding prey. In some forms the 'thumb' and big toes are lost.

The flesh-eating forms were dominant in the first epoch of the Tertiary, the Eocene. These are referred to as archaic carnivores or ceodonts, and are now all extinct. The range in size was great, some comparable to modern weasels while others evolved quite large sizes. The adaptive radiation which took place in this hunting order is wide. For example, the seals, sea-lions and walruses are marine carnivores that have existed since the Miocene. The bears were a Miocene offshoot from the dog stock, both groups are comparatively primitive carnivores and associated with the typical small carnivores, the weasel family. The most modified are the civets and mongooses (Viverridae), the hyenas and the true cats (Felidae).

A calicothere

Hoofed mammals

There are some twenty living species of herbivorous ungulates. During the Palaeocene a number of animals abandoned the insectivorous habit and began to eat plants. The early ungulates are known as the condylarths and resemble the early carnivores, with a long body and limbs short and primitive in structure. However, the cheek teeth evolved for chewing vegetable food and although all the toes were present each was capped by a small hoof.

Living odd-toed ungulates, *perissodactyls*, include the horses, rhinoceroses and the tapirs. The foot symmetry has the axis through the middle toe and there has been a tendency for reduction to a single toe.

Of the early ungulate forms, the extinct horse-like titanotheres began in the Eocene and rapidly evolved to gigantic sizes. Being slow and cumbersome they fell prey to the larger carnivores of their day.

Tapir

Rhinoceros

Not uncommon in the Tertiary were the calicotheres, similar to horses as both had titanothere teeth, suitable only for soft plant food. Their toes ended in huge claws.

The only living representative of the tapirs are confined to the tropics of South America and the Malay region. In the Tertiary they were widespread throughout North America and Europe, disappearing from these regions with the cold of the Pleistocene.

Rhinoceroses are today confined to the Old World tropics. The early forms of the Eocene were slim, small, horse-like in build and were without horns.

A titanothere

Pleistocene – Recent — *Equus*

Pliocene — *Pliohippus*

Miocene — *Merychippus*

Oligocene — *Mesohippus*

Eocene — *Hyracotherium*

Evolution of the horse

Probably the most widely known of all evolutionary examples is that of the horse. North American Tertiary deposits contain a very good fossil series from grasslands, where most of the horse evolution took place. The earliest horses were whippet-sized and known from the Eocene beginnings. From the chart illustrating their subsequent evolutionary trends we can observe four main changes. These involve an overall increase in size, lengthening of the limbs and modification of foot structure, changes in the skull and teeth, and an increase in both the relative size and complexity of the brain.

Whereas *Eohippus* stood 12 inches at the shoulder, the modern carthorse may stand well over 5 feet. Although there are obvious advantages in an increase in size there are also several problems. The great increase in weight involves increased stresses on the limbs which must therefore evolve strengthened bones. Also feeding habits become more complex as larger amounts of food are required. The modern horse is by no means just an enlarged edition of *Eohippus*. The whole animal has undergone drastic changes in various structures, on the whole related to the overall increase in size.

The teeth changed in design and size due to the increased bulk and the change in feeding habit. They gradually increased in length, surface area and structural efficiency. In the Miocene they show radical change probably due to the horse becoming a grazer rather than a browser in feeding habits, related to the change in environment at this time. There is abundant fossil grass seed evidence that during this period there was a transition from the hardwood forests to open forests and prairie.

With the change in limbs and the increased stress placed upon them, there was a reduction in toes, from a 'pad-footed' type, to a 'spring-footed' type. The reduction to a single hoof also reflects the change to life on the plains for the horse. The single hoof would have been highly unsuitable for a forest environment but ideal for high running speeds on the firmer ground of the prairies.

Development of limbs, skulls and form during evolution of the horse. Note the reduction in the number of toes which allowed greater speed over hard ground. The reconstructions are to scale.

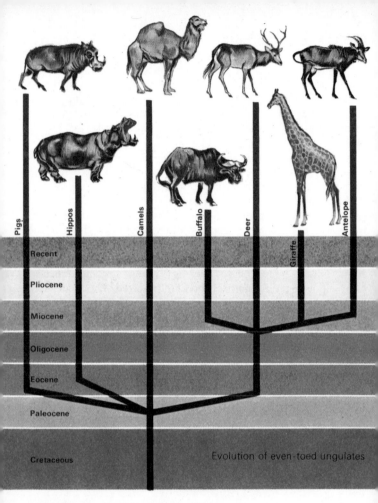

Evolution of even-toed ungulates

Even-toed ungulates

Even-toed ungulates or *artiodactyls* have for many thousands of years been of extreme importance to man, and to the carnivores whose lives depended on them. Although present at the beginning of the Eocene, it was not until later that they became a prominent and diversified group. There seems to have been a distinct division into pigs and ruminants.

Pigs and their relatives are, in many ways, the most primitive of living even-toed ungulates. They are four-toed animals although the side toes are reduced. The limbs have not elongated so speed has not developed.

The hippopotamus is distantly related to the swine found in the Pleistocene in regions as far removed as England and China. Only two species survive today.

The success of the ruminants is undoubtedly largely due to the four-chambered stomach which allows the animal to digest tough grasses. The complex arrangement allows ruminants to gulp large amounts of food quickly, retire to a safe spot and regurgitate it from the first two chambers, to be chewed and digested at leisure. The ruminants were the last important mammalian group to evolve, most of them branching off in the Eocene. The antlered deer, giraffe and okapi, the bovids (cattle, sheep and goats) have more complex stomachs still.

Camels and llamas were the earliest group to evolve and have had a long distinct history. North America was their original home in the Oligocene and Miocene but the Pleistocene llamas had become established in South America and the true camel had migrated to Asia.

One might query the success of the even- over the odd-toed ungulates. The teeth, the brains and the feet played an important part, but probably the development of the stomach was the true reason for the marvellous success of the artiodactyls.

Toe variation among even-toed ungulates

Hippopotamus   Deer   Camel

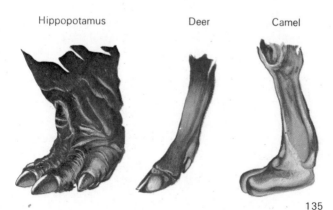

Elephants and relatives

Fossil evidence shows that in the past the elephants have been adapted to all varying types of climate.

The earliest form traced is that of *Moeritherium*, a small creature showing few of the elephantine characteristics we imagine today. There was no trunk (perhaps a pig-like snout), and all teeth were present, though signs of an enlarged upper and lower pair were visible. Certain authorities debate whether or not it should belong to the elephant group as it had these primitive characteristics. Gradually an increase in size took place, the jaws shortened, tusks evolved, the upper ones becoming free and curved, the lower ones eventually reducing, and a highly flexible trunk formed. The two modern elephants and the mammoths derived from the mastodonts of Miocene stock. The evolutionary pattern is quite complex and the sequences are mostly guesswork.

The early mammoths originated in Asia and soon spread over Europe and descendants migrated to North America where they gave rise to gigantic forms. The Woolly Mammoth (*Mammuthus orimegenius*), adapted to steppe country, and was quite common during the last glacial period. The true appearance is well known due to discoveries of whole specimens in the ice of Siberia.

Although there are numerous other differences between the modern African Elephant (*Loxodonta africana*) and the Indian Elephant (*Elephas maximus*), the usual identification is that the African has large, and the Indian small ears.

The hyraxes (Procaviidae) are thought to be relatives of the elephant having branched from the elephant development probably in mid-Palaeocene times. Superficially they resemble rabbits, but internally there are certain distinguishable links with the elephant.

The sea cows (*Sirenia*) of today include the dugong of the Indian Ocean and the Red Sea and the manatee of the shores of the tropical Atlantic. They are aquatic, browsing on coastal vegetation and in the Tertiary were quite widespread. Resemblance to older fossils link the *Sirenia* unmistakably to the elephants.

Evolution of the elephant

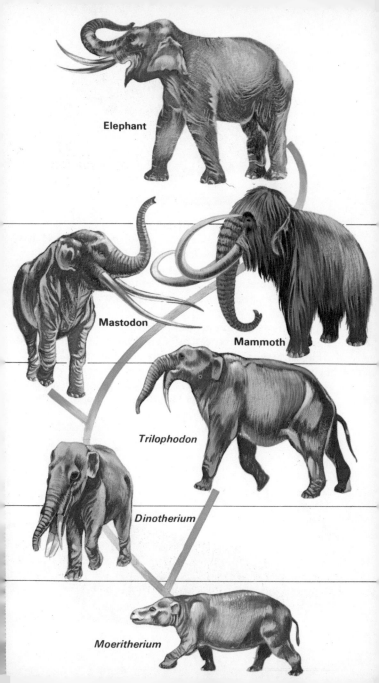

Elephant

Mastodon

Mammoth

Trilophodon

Dinotherium

Moeritherium

Rat

The diversity of mammals

Rodents are the most numerous of all mammals, and have migrated during their evolution to live in nearly every corner of the Earth, from the Arctic to the Tropics. Although they have a small brain, they have proved to be the most successful of mammals, because of their adaptability and extremely rapid breeding.

The whales are almost completely adapted to an aquatic life. Although their ancestry is uncertain, they are thought to have diverged from early entherian stock. Fish-like in shape, their internal adaption shows the most extreme specializations of any mammal. Vestigial hind limbs of ancestral terrestrial forms can be found in whales, not connected to the spine. The forelimbs are flipper-like.

Although there is no intermediate fossil evidence, bats are

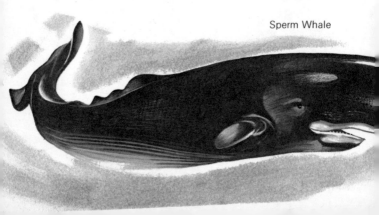

Sperm Whale

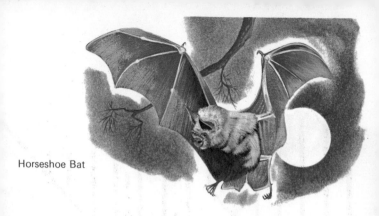

Horseshoe Bat

thought to have branched from a primitive stock of the shrew-like insectivores. There existed several different species in the Eocene, and today there are some 750 species.

The armadillos, anteaters, and sloths, collectively known as the edentates, are peculiar to South America. The anteaters feed on the termites of tropical regions using strong claws to rip nests apart.

The sloths hold on to the branches of trees, using long curved claws, hanging in the characteristic upside down position. Tree-sloth fossils are very rare, but whole skeletons have been found of the extinct giant ground sloths. *Megatherium* of the Pleistocene period grew to elephantine proportions.

The armadillos are protected by bony plates covered with horn. The giant armadillo, *Glyptodont*, of the Miocene, shows that some grew to over 10 feet long.

Sloth

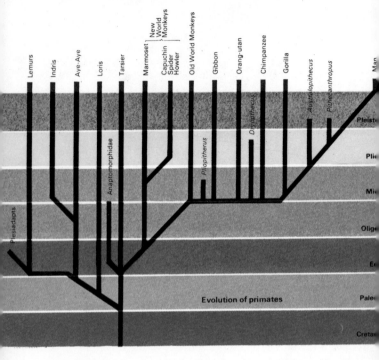

Evolution of primates

Primates

The order Primates includes apes, monkeys, tree-shrews, lemurs, bush-babies and tarsiers. Fossil evidence to link the families is rather scarce due to the fact that most are arboreal. The limbs of the primates are quite primitive when compared to other mammalian specializations. In most lower forms the primitively long tail is retained as a balancing organ, but in some higher forms it becomes reduced or lost all together. During evolution of the primates changes to improve tree-dwelling life are observed; the sense of smell degenerates as vision improves; the brain centres develop to give excellent muscular and nervous co-ordination.

The structure of the tree-shrews shows them to be inter-mediates between insectivores and primates, but their brain is much more advanced and complex. The tree-shrews have changed little since Palaeocene times when they first branched from the insectivores.

From the tree-shrews arose the rather more advanced lemurs and tarsiers. All are arboreal, feeding on fruit and insects, and now survive mainly in restricted areas. The lemurs are found only in Madagascar, the lorises in Asia and the galago or bush-babies and potto inhabit areas of Africa. In early times they were all quite widespread but competition from monkeys, and the radiating flesh-eating carnivores reduced their numbers. The curious Aye-aye has claws on all toes except the big toe and procumbant incisors like a rodent, these specializations allowing it to obtain bark-dwelling insects. The lorises of India move in a deliberate manner being unable to jump, as do the pottos of Africa, but the galagos are able to jump magnificently with elongated back legs.

The one living form of the tarsiers, *Tarsius,* is of the East Indies and Philippines. Fossil forms date from the Palaeocene and twenty five genera have been identified from North American and European deposits. Too specialized to be on the direct line of descent of the higher primates, *Tarsius* does however show links. The neck is very mobile as the large, forward-directing, nocturnal eyes move very little. The ankle bones of the foot are greatly elongated lengthening the foot and giving *Tarsius* powerful long jumping hind legs.

Nocturnal tree-climbers, the Aye-aye (*top*) and Potto (*bottom*) The Potto is the African equivalent of the Loris.

Mandrill

Patas

Old world monkeys

These are the monkeys of Asia and Africa evidence for which dates back to the Oligocene. The more primitive forms are arboreal but the baboons and relatives show a tendency towards a ground-dwelling life. Two groups are distinguished today; one group has cheek pouches and the other has a complicated type of stomach (similar to ruminants).

The cheek-pouched groups are frequently found in zoos. The macaques, Rhesus Monkey and relatives are specialized, with less arboreal habits, longer grinding teeth and somewhat longer faces. The baboons, mandrills and drills have a characteristic dog-like snout due to the elongated teeth evolved for a herbivorous diet.

The langurs of Asia are examples of the second group and one member is the Proboscis Monkey with its wonderfully elongated pendant nose, still an evolutionary mystery.

Gelada

Rhesus

Marmoset

Uakari

New world monkeys

The South American monkeys are known as flat-nosed monkeys because their nostrils are widely separated and face more sideways than downwards and forwards.

The smallest forms are the small colourful marmosets, squirrel-like, with thick long fur, bushy tails, and claws. The marmosets and tamarins, unlike most other anthropoids, give birth to two or three young at a time.

The cebids are the other South American monkey group. The Spider Monkey is a typical member with a prehensile tail. Capuchin monkeys will eat fruit and insects whereas others feed only on fruit. The uakaris are rather bizarre cebids with bald heads and long brown hair.

In all monkeys the eyes are large and facing upward, giving binocular vision. An increased brain size allows for a quite high level of learning and intelligence.

Spider Monkey

Capuchin

Apes

Today the only representatives are the gibbons, Orangutan, Chimpanzee and Gorilla. All are more advanced than monkeys, notably in the larger and greater complexity of the brain, and the different tooth pattern. Due to their large size apes can no longer run along the tops of branches in monkey fashion. They have become great movers by swinging with their arms which are longer than their legs and provided with very powerful muscles. These changes have meant changes in the body structure, the chest has widened, the neck and limbs lengthened, and the head enlarged.

There are close anatomical and physiological resemlances between the apes and modern man. There is no structure in the Gorilla that cannot be found in man. The relative development and size of certain body parts are quantitatively different, the most important ones associated with locomotion and brain size. This does not mean that man evolved from the apes and the two groups probably diverged from a primitive

Gibbon (*top*) and
Chimpanzee (*bottom*)

basic stock in the early Miocene.

Fossil evidence of the apes is not common, but examples show that these primates were widely distributed throughout Europe, Asia and Africa during the middle and latter portions of the Cenozoic.

Parapithecus is described by some authorities as an early Cercopithecoid monkey, whereas others regard it as an early ape. Perhaps it was common to both, one cannot at present state precisely where it should be placed. In the same Egyptian beds in the lower Oligocene sediments, a jaw was identified as that of a small primitive ape, *Propliopithecus*.

The earliest apes to appear after *Propliopithecus* were *Limnopithecus* and *Proconsul* from the lower Miocene sediments of Africa.

There seem to have been apes living in Asia during the pleistocene epoch that were larger than the gorillas. Today the gorilla and chimpanzee inhabit Central Africa, the Orang-utan Borneo and Sumatra, and the gibbons the tropical forests of Indonesia and South East Asia.

Orang-utan (*top*) and Gorilla (*bottom*)

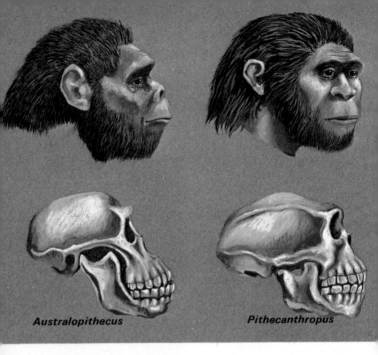

Australopithecus

Pithecanthropus

From apes to man

Apart from *Kenyapithecus* and *Ramapithecus*, which might possibly be ancestral hominids, the first men lived over a million years ago, during the early part of the Pleistocene. These are the 'southern ape men' or *australopithecines* of Africa, and it appears that it was about this time that the hominids were beginning to walk upright. At first they were regarded as apes but their erect position, hip bone structure and dentition led to their reclassification as hominids. The brain was gorilla-sized, but the body much smaller, so relatively the brain was much larger than a gorilla's.

First evidence found was at Taung, Bechuanaland, in 1924. More recently two different australopithecines have been discovered in the lower deposits at the Olduvai Gorge, Tanzania. At a lower level in the gorge, a larger brained *pre-Zinjanthropus* was found and may be more advanced than the other australopithecines.

During the middle Pleistocene there was a gradual replace-

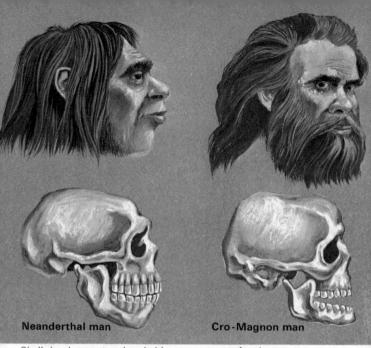

Neanderthal man   **Cro-Magnon man**

Skull development and probable appearances of early men

ment of the australopithecines by the larger-brained true men. Remains have now been discovered in Java, Peking, and Algeria, and at Olduvai Gorge. All had very heavy brow-ridges and the vault of the skull was much lower than that found in modern man. The skull changed gradually over the next hundreds of thousands of years, the top enlarging to make room for the greatly enlarged brain. The hominids learnt to make fire and cook, thus making powerful jaws and huge meat-tearing teeth unnecessary. This meant that well-developed brow-ridges supporting muscles working the jaws gradually disappeared and a high forehead developed. The Neanderthal men (*Homo neanderthalensis*) were replaced by modern men (*Homo sapiens*) some 40,000 years ago.

Due to man's capabilities of thinking and logical reasoning, he has been able to radiate to fill practically every corner of the Earth. He now can control his own evolution and more or less decide the future of all other forms of life.

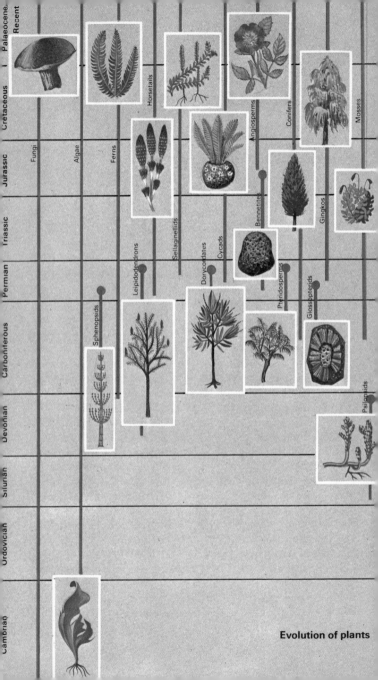

Evolution of plants

Palaeocene, Recent
Cretaceous
Jurassic
Triassic
Permian
Carboniferous
Devonian
Silurian
Ordovician
Cambrian

Fungi
Algae
Ferns
Horsetails
Angiosperms
Conifers
Mosses
Sellaginellids
Bennettites
Gingkos
Lepidodendrons
Dorycordates
Cycads
Pteridosperms
Glossopterids
Sphenopsids
Psilophids

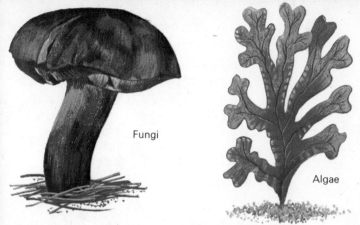

Fungi

Algae

Representatives of the primitive plant group Thallophyta

Evolution of plant life

As plant and animal life evolved in water, the properties of this medium imposed certain restrictions on the evolutionary possibilities of organisms living there. The first plants were unicellular or multicellular organisms with simple structure, without differentiation into stems, leaves, and roots. They are known as the **Thallophyta** and are represented by algae of various kinds including seaweeds and stoneworts.

Plants evolved no further until they colonized the land. There is evidence in the Cambrian period of spores from land plants but in the Devonian the land was already partially covered by Thallophyte plants. At first the thallophytes lived in shallow water. So that the plants could stand erect, resist desiccation, take in atmospheric carbon dioxide and oxygen, and obtain food materials, the thallophytes evolved a stiff outer layer perforated with stomata. The roots absorbed water and the mineral salts dissolved in it from the soil. The method of reproduction also had to adapt. In water the thallophytes reproduce sexually by eggs and sperms, or asexually by spores, and in order to reach the egg and fertilize it, the sperm must swim through the water. On land therefore, a thin film of moisture is required to ensure fertilization.

The **Psilophyta** were the simplest land plants known – well-preserved remains have been found in the Devonian period.

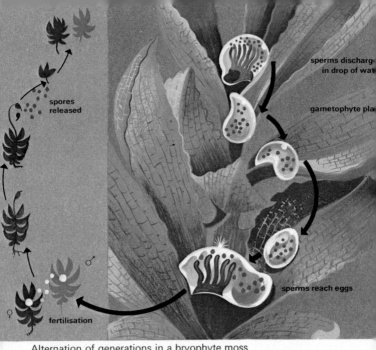

Labels on image: sperms discharged in drop of water; gametophyte plant; sperms reach eggs; spores released; fertilisation; ♂; ♀

Alternation of generations in a bryophyte moss

One *Rhynia major*, consisted of a stem continuous with an underground stem or rhizome which was not differentiated into a root but bore absorbent root-hairs. The stem branched and at the end of some branches were spores in groups of four, as in ferns. Stomata, small openings that allow the flow of air, are visible in the stems, just as in all land plants. Rudimentary leaves were found in a related plant, *Asteroxylon mackiei*. These Devonian plants evolved from the marine algae. Evidence to establish that the Psilophyta are directly ancestral to any higher land plants is lacking. However, they do show the appearance of the plants that first colonized dry land. They could live only in damp places because a film of moisture is necessary to enable the sperms to pass to the egg. In the classification of plants the method of reproduction usually positions the individual species.

The **Bryophyta** (mosses and liverworts) have a fairly simple cellular structure which restricts their size. They have not evolved very far and have developed a reproductive method known as alternation of generations. The moss plant is known as the *gametophyte* as it produces the eggs and sperms. The fusion of these give rise to the small *sporophyte* which lives parasitically on the moss leaf. It grows and liberates spores which germinate into moss plants. As the gametophyte is haploid it is less well provided with possible variants in the genetic make up.

In the more advanced **Pteridophyta** (ferns, horsetails and selaginellids) the plant is the sporophyte, while the gametophyte is minute. The sporophyte has the diploid condition of chromosomes and so is well equipped to provide variants. Thus natural selection has been able to work and the results are evident when we look at the plants themselves. They have evolved 'vessels' running through the root and stem systems for transport of water and food substances, and are thus sometimes called the 'vascular cryprograms'. The life cycle diagram illustrates the alternation of generations in a fern but the sequence is the same for all pteridophytes.

Alternation of generations in a pteridophyte fern

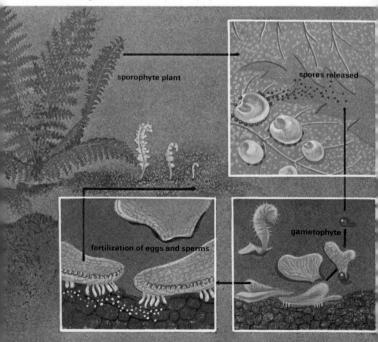

Cryptograms and phanerograms

The thallophytes, psilophytes, bryophytes and pteridophytes are grouped together and called *cryptograms* ('hidden marriage') as they do not produce by visible seeds, but spores. Selaginellids, such as *Lycopodium*, show a transition stage in the evolution of seeds. They have two kinds of spores; large 'megaspores' develop into female prothalli bearing egg-producing *archegonia*, and small 'microspores' which develop into male sperm-producing *antheridia*.

Certain cryptograms evolved to produce types which suppress the gametophyte generation completely. It does not exist as a separate structure, but is represented by a few cells and nuclei only when the spores are produced. The pollen grain is the microspore and one of its nuclei develops to be the sperm. The egg-sac is the megaspore with the egg inside it.

The pollen-grains are shed into the air, and usually transported by the wind. The megaspore is not shed but is retained within its spore-case or ovule. The ovule is exposed to the air and at a certain part of its coat allows a pollen-grain to come into contact with the megaspore. The pollen-grain is able to penetrate the egg by growing a pollen-tube and a nucleus, acting as the sperm, passes down as it grows until the egg is reached. No film of water or moisture has been required and it is this advancement in the reproductive mechanism that enabled plants to spread and colonize dry land.

Once fertilized the egg becomes an embryo and develops within the ovule which forms the seed. If conditions and environment are suitable the seed's embryo will germinate and produce a seedling and eventually a new plant. In this method of reproduction, seeds are formed and eventually expelled, and the plants are known as *phanerograms* ('conspicuous marriage').

'Gymnosperms' (naked seeds) are the most primitive seed-bearing plants. Among the living members are the cycads, the ginkgo, the conifers and the Pteridosperms and Cordaitae. True fossilized gymnospermous wood in mid-Devonian rocks has a structure as highly organized as the secondary wood of living conifers. Several groups seem to have evolved during the Triassic and Jurassic to maximum development, and then disappeared in succession. Jurassic fossils show relatable characters to the living *Araucaria* (monkey-puzzle tree).

The gymnosperms bear male and female cones. The female cones are formed from the grouping of leaves bearing the ovules, and the male cones from the grouping of the stamens.

Flowering plants

About a quarter of a million living species are found in the group of flowering plants. Here, evolved the complete protection of the ovule by its enclosure inside an 'ovary', a chamber formed from the carpellary leaves that bear the ovules. From the ovary, projecting upwards, is a thin cylindrical style. The end known as the stigma is sticky so that pollen grains touching it will adhere. The pollen grains grow down the stigma shaft and the first to reach the ovule containing the egg, fertilizes the egg. Plants with this type of seed and reproduction are all called 'angiosperms' and include all plants which have true flowers. They are so highly organized it is impossible to decide which is the simplest form of flowering plant.

Reproduction in flowering plants

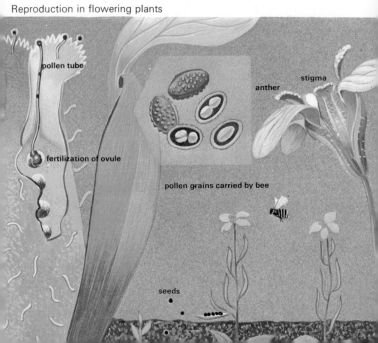

pollen tube

fertilization of ovule

anther

stigma

pollen grains carried by bee

seeds

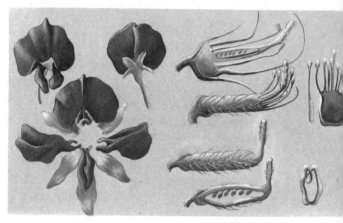

Anatomy of a flower

From fossil evidence it seems that angiosperms developed in the Cretaceous. Certain lines gave rise to the grasses which were very important in the evolution of the hoofed grazers such as horses, deer, antelopes and cattle.

Several factors are evidently involved in the success of the angiosperms. Increasing aridity in late Jurassic and early Cretaceous times favoured this group. Furthermore, the seas retreated exposing barren land which was quickly colonized. The conifers had depended on wind pollination and seed dispersal. However, within the angiosperms various forms of pollination and dispersal evolved, effected by insects, birds and mammals (see page 65). Although each angiosperm flower has male and female parts there are always elaborate mechanisms to prevent self-fertilization and therefore ensure cross-pollination.

The Angiosperms can be divided into two main groups, the monocotyledons and the dicotyledons. The seed contains the 'cotyledon' which is a leaf-like structure responsible for nourishing the germinating embryo. Monocotyledons have one cotyledon, dicotyledons have two.

The *monocotyledons* of the angiosperms include grasses, sedges, lilies, palms, pineapples, orchids, and crocuses. They have a single seedling leaf which grows to produce a plant

Examples of monocotyledons

with leaves showing parallel venation. Another distinguishing point is that no thickening of the stems takes place in the form of bark, wood or pith.

The *dicotyledons* include the vast majority of flowering plants such as the vegetables (excluding the onion), fruits, shrubs, deciduous trees and the majority of garden plants (excluding the lilies, orchids, and tulips).

Examples of dicotyledons

BOOKS TO READ

The following books are recommended for further general reading and are usually obtainable from bookshops and public libraries.

The Theory of Evolution by J. Maynard Smith. Penguin (Pelican), Harmondsworth, 1958.
Evolution of the Vertebrates by E. H. Colbert. Wiley Interscience.
The Voyage of the Beagle by C. Darwin. Dent, London, 1961.
Fossils, a Guide to Prehistoric Life by F. H. T. Rhodes, H. S. Zim and P. R. Shaffer. Golden Press, New York, 1963.
Dinosaurs, their Discovery and their World by E. H. Colbert. Hutchinson, London, 1962.

The following more specialized books are published by the British Museum (Natural History).

The Succession of Life through Geological Time by K. P. Oakley and H. M. Muir Wood. 1964.
Fossil Amphibians and Reptiles by W. E. Swinton. 1962.
Fossil Birds by W. E. Swinton. 1958.
Dinosaurs by W. E. Swinton. 1964.
History of the Primates by Sir W. E. Le Gros Clark. 1962.

INDEX

SOME OTHER TITLES IN THIS SERIES

Natural History

The Animal Kingdom
Animals of Australia
& New Zealand
Bird Behaviour
Birds of Prey
Fishes of the World
Fossil Man
A Guide to the Seashore

Life in the Sea
Mammals of the World
Natural History Collecting
The Plant Kingdom
Prehistoric Animals
Snakes of the World
Wild Cats

Gardening

Chrysanthemums
Garden Flowers

Garden Shrubs
Roses

Popular Science

Astronomy
Atomic Energy
Computers at Work
Electronics

Mathematics
Microscopes & Microscopic Life
The Weather Guide

Arts

Architecture
Jewellery

Porcelain
Victoriana

General Information

Flags
Guns
Military Uniforms
Rockets & Missiles
Sailing

Sailing Ships & Sailing Craft
Sea Fishing
Trains
Warships

Domestic Animals and Pets

Budgerigars
Cats
Dog Care

Dogs
Horses & Ponies
Pets for Children

Domestic Science

Flower Arranging

History & Mythology

Discovery of
  Africa
  North America
  The American West
  Japan

Myths & Legends of
  Africa
  Ancient Egypt
  Ancient Greece
  The South Seas